ON THE

THEORY AND CALCULATION

OF

CONTINUOUS BRIDGES

BY

MANSFIELD MERRIMAN, C.E., Ph.D.,

INSTRUCTOR IN CIVIL ENGINEERING IN THE SHEFFIELD SCIENTIFIC SCHOOL OF YALE COLLEGE.

ILLUSTRATED.

NEW YORK:
D. VAN NOSTRAND, PUBLISHER,
23 MURRAY AND 27 WARREN STREET.
1876.

PREFACE.

The following pages are divided into three chapters. The first presents by way of introduction some of the elementary principles of continuous girders, and the fundamental ideas relating to the calculation of strains. The second gives the theory of flexure as applied to the continuous truss of constant cross section, and exhibits it in formulæ (I) to (VI), ready for application to any particular case ; and the third gives an example of the computation of strains in a continuous truss of five unequal spans, with some useful hints concerning the practical building of such bridges.

The theory of flexure indicates that, by the use of continuous instead of single span bridges, a saving in material of from twenty to forty per cent. may be effected. It is easy indeed to say that this advantage will be entirely swallowed up by the effect of changes of temperature, increased labor of erection, or additional cost of workmanship, but by no amount of reasoning can such disadvantages be estimated. Theory indicates a large saving, whether or not it can be realized. may only be determined by trial. Other nations have built and are building continuous bridges, and their experience has not yet shown that the system is inferior to that of single spans. The interest now prevailing among American engineers in the subject, and the fact that at some recent bridge lettings plans have been offered for a continuous structure, seem to indicate that the system will also be tried here.

This little book may then perhaps be of value to bridge engineers, as well as to students in general.

M. M.

New Haven, Conn., July 10, 1876.

ON THE

THEORY AND CALCULATION OF CONTINUOUS BRIDGES.

WHEN a straight bridge consists of several spans, each entirely independent of the others, it is said to be composed of *simple girders*. If, on the other hand, it consists of a single truss extending from one abutment to the other without any disconnection of parts over the piers it is called a *continuous girder*. A load placed upon any span of a continuous beam influences, to some extent, each of the other spans, and hence its complete theory is much more complex than that of the simple one. This very complexity however has rendered the subject an attractive one to mathematicians, who, pursuing science for science's sake, have investigated the laws of equilibrium which govern it. These laws with the many beautiful consequences attending them form one of the most interesting chapters of mathematical analysis, and as such have interest and value independent of their application in engineering art.

It is the object of the present paper to present in as simple a form as possible some of the main principles and laws most needed by the engineer, and to illustrate their application as fully as space will permit to the practical designing of continuous bridges.

Chapter 1.

The first point to be observed in considering either a simple or continuous girder is that all the exterior forces which act upon it are in equilibrium. The exterior forces embrace the weight of the girder and the loads upon it which act downward, and the pressures or reactions of the supports which act upward. In order that these may be in equilibrium, it is necessary that *the sum of the reactions of all the supports must be equal to the total weight of the girder and its load.*

Thus, if a simple girder of uniform section and weight rest at its ends upon two supports, the reaction of each support will be one-half the weight. Exactly in the center between the two supports or abutments, let us suppose a pier to be placed just touching, but not pressing against the beam, which, at that point, has a deflection below a straight line joining the two abutments. Then the condition of things is in no way altered, for the weight being W, each abutment reacts with a force $\frac{1}{2}$ W, while the pier bears no load. Raise now the pier so as to lift the girder above the line of deflection and it receives a part of the weight W, while the reactions of the abutments become less than $\frac{1}{2}$ W. If the pier be raised higher and higher, it will at length lift the girder entirely from the abutments and bear itself the whole load W. In every position, however, the sum of the reactions of the three supports is equal to the total load. For example, when the three are on the same level it may be shown that the reaction of each abutment is $\frac{3}{16}$ W, and that of the pier $\frac{10}{16}$ W.

This illustration shows also that *small differences of level in the supports occurring after the erection of a bridge cause large variations in the reactions of its supports and in the strains in its several parts.* A simple girder having a deflection of one inch, would, if raised one and three-fifth inches at the center, be entirely lifted from the abutments. In the first case the upper fiber would be in compression, the lower in tension; in the second case, the upper would be in tension, the lower in compression. If the center were raised only one inch, the reversal would be only partial, the upper fiber becoming subject to tension for a short distance on each side of the middle. This fact often used as an argument against continuous bridges, is really an objection only when the piers are liable to settle after erection. Differences of level, previously existing, do not act prejudicial when the bridge is built upon the piers, and with a profile corresponding to them.

The mathematical theory of the continuous girder enables its reactions and internal strains to be found for any assumed levels of the supports, provided only that the differences of level are very small compared with the length of the spans. However interesting such investigations may be in themselves, they are of little importance in practice, since it has been shown that when all the points of support are on the same level, the greatest economy of material results.* In all that follows, then, we shall regard the girder as resting on level supports, or, what is the same thing, that it was built with a profile corresponding to that of the piers.

The loads upon a bridge and the reac-

* Weyrauch; *Theorie der continuirlichen Träger*, p. 120. Winkler; *Lehre der Elasticität*, p. 155.

tions of the supports are *external forces.* The equilibrium between them is maintained by means of *internal forces*, which, in a framed truss, are transmitted longitudinally along the pieces as strains of tension and compression. *When all the external forces are known, these internal forces or strains can be readily found.* This very important point we shall now proceed to illustrate.

Fig. 1 represents a portion of a con-

Fig. 1.

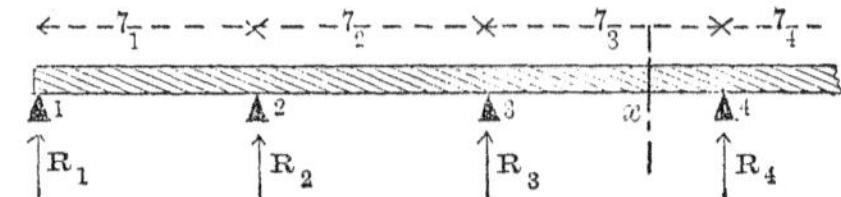

tinuous girder; the first span on the left is called l_1, which also represents its length, the second l_2, the third l_3, etc.; in like manner, the supports beginning on the left are designated by the indices 1, 2, 3, etc., and their reactions by R_1, R_2, R_3, etc. Let the load per linear unit be w, supposed uniformly distributed, then the weight of the first span will be $w\,l_1$, of the second $w\,l_2$, of the third $w\,l_3$, etc. If there are four spans the total weight will be

$$w\,l_1 + w\,l_2 + w\,l_3 + w\,l_4$$

and from the fundamental idea of equilibrium, we must have the equation

$$R_1 + R_2 + R_3 + R_4 + R_5 = w\,(l_1 + l_2 + l_4 + l_4)$$

Each of the reactions is then a fractional part of the total load, and by methods hereafter to be explained, their values may be readily computed, whatever be the number of spans. Granting for the present that they may be found, let us inquire how we may obtain the internal forces or strains in any part of the girder.

In the span l_3 let a vertical plane be passed, cutting the beam at a point whose distance from the support 3 is x. All the internal forces acting in this section may be considered as resolved into two components, one vertical and the other horizontal. The sum of all the vertical components is a force which prevents the two parts of the beam from shearing asunder, and is called the *shear* for that section ; the horizontal components acting in parallel planes are the resisting strains of tension and compression in the horizontal fibers, and the sum of their moments with reference to any point in the section is called the *moment of resistance*, or simply the *moment* for that section. The internal strains in any section are thus completely represented by the shear and moment. For example, if the girder in Fig. 1 be a framed truss of which Fig. 2 represents the span l_3

Fig. 2.

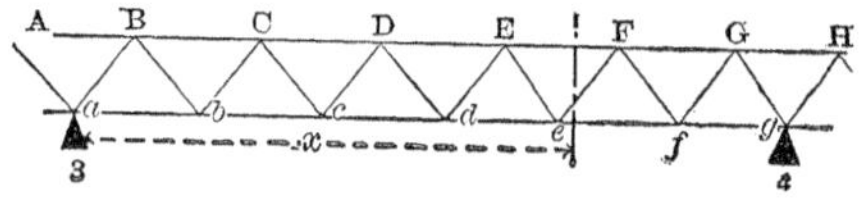

enlarged, and the section be passed cutting the three pieces E F, F *e*, and *e f*, the vertical components of the chord strains will be zero, and that of the diagonal strain *e* F will be the shear. Hence, if the shear be known, and the angle included between the vertical and a diagonal be θ, we have only to multiply the shear by *sec.* θ to find the strain in the diagonal. Again, let the section be moved to the left so as to pass through the point *e*, and let that point be taken as the center of moments. Then the moment of resistance will be the moment of the chord strain E F. Hence, if that moment be known, we have only to divide it by the depth of the truss to get the strain in E F.

The internal shear and moment at any section are easily found from the fundamental conditions of statical equilibrium. The shear being an internal vertical force is the resultant of the exterior vertical forces on either side of the section. The exterior forces on the left of the section, for instance, have for a resultant their algebraic sum; considering the upward forces as positive, and the downward ones as negative we have from Fig. 1, their sum

$$S = R_1 - w\, l_1 + R_2 - w\, l_2 + R_3 - w\, x$$

as the expression for the shear in the section x. To get the internal moment for the same section, we have only to consider in like manner that it is equal to the sum of the moments of all the exterior forces on either side of the section, for if otherwise, there would be a tendency to rotation. The moment of the force R_1 with reference to x is $R_1 (l_1+l_2+x)$, of R_2 is $R_2 (l_2+x)$, of the load $w\,l_1$, is $w\,l_1 (\frac{1}{2}l_1+l_2+x)$, etc. Thus from a mere inspection of Fig. 1 we write the value of the moment M, regarding those moments as positive which cause a tensile strain in the upper fiber at x, and those as negative which cause a compressive one. The expression is

$$M = -R_1 (l_1+l_2+x) + w\,l_1 (\tfrac{1}{2}\,l_1+l_2+x) - R_2 (l_2+x) + w\,l_2 (\tfrac{1}{2}\,l_2+x) - R_3\,x + wx \cdot \tfrac{1}{2}\,x$$

Now in these expressions for the internal shear S and the moment M at any point x, the lengths l_1, l_2, x are given by the conditions of the case in hand, and the same is true of the load per linear unit w. Hence the shear and moment, and consequently the internal strains are easily obtained as soon as the reactions of the supports are known. We shall hereafter give methods by which the reactions may be readily determined.

By the same reasoning if we pass a section in the span l_2, (Fig. 1) at a point whose distance from the support 2 is x, the shear S and the moment M for that section will be

$$S = R_1 - w\,l_1 + R_2 - w\,x$$
$$M = -R_1\,(l_1 + x) + w\,l_1\,(\tfrac{1}{2}\,l_1 + x) - R_2\,x + \frac{w\,x^2}{2}$$

In a simple girder whose length is l_1, and each of whose reactions is R or $\frac{1}{2}\,w\,l$, the shear and moment for any section x will be

$$S = R - w\,x = w\,(\tfrac{1}{2}\,l - x)$$
$$M = -R\,x + \frac{w\,x^2}{2} = \frac{w}{2}\,(-l\,x + x^2)$$

When the number of spans is large, the expressions for the shear and moment as above deduced become long and involve much arithmetical computation. We are, however, fortunately able to place them under a much simpler form. First they may be written thus

$$S=(R_1-w\,l_1+R_2-w\,l_2+R_3)-w\,x$$
$$M=[-R_1\,(l_1+l_2)+w\,l_1\,(\tfrac{1}{2}\,l_1+l_2)-R_2\,l+\tfrac{1}{2}\,w\,l^2]-(R_1-w\,l_1+R_2-w\,l_2+R_3)x+\tfrac{1}{2}\,w\,x^2$$

Now $(R_1-w\,l_1\;+R_2-w\,l_2\;+R_3)$ is the shear in the span l_3 at a point infinitely near to the support 3 ; let this be called S_3. Also the quantity enclosed in [ ] in the second equation is the moment of the exterior forces with reference to the point 3 ; let this be called M_3. Then the equations become

$$S=S_3-w\,x$$
$$M=M_3-S_3\,x+\tfrac{1}{2}\,wx^2$$

Therefore *the internal shear and moment at any section can immediately be found,* without the necessity of determining the reactions, *provided we know the shear and the moment for the preceding support.* This method, due to Clapeyron, of using the moment at the supports instead of the reactions greatly simplifies the numerical computations of a continuous truss. We designate the moment at 3 by M_3, the reaction being R_3, or the sum of the shear S_3, in the span l_3 and of the shear S'_2 in the span l_2 both infinitely near to the support 3. In like manner the moments at the supports 2 and 4, will be designated by M_2 and M , the shears just to the right of those points by S_2 and S_4, and those to the left by S'_1, and S'_3. In general for any support whose index is n, we have (Fig. 3)

Fig. 3.

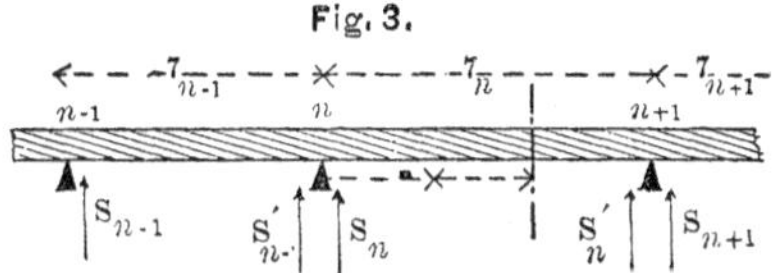

on the left, the span l_{n-1}, on the right the span l_n ; the shear infinitely near to n on the left is S'_{n-1}, on the right S_n ; the sum of S'_{n-1}, and S_n makes the reaction R_n ; and the moment over the support is M_n. If the load be uniform and equal to w per linear unit, the internal shear S and moment M for any section x are given by

$$S = S_n - w\,x$$
$$M = M_n - S_n\,x + \tfrac{1}{2}\,w\,x^2$$

To find then the internal strains in the diagonals of a continuous truss due to dead load only, we have to pass a section cutting each diagonal and find the value of S, this is the shear which the strain in the diagonal must resist and multiplied by *sec.* θ (θ being the inclination of the diagonal to the vertical,) it gives the required strain. To find the strains in the upper chord we have to take the lower chord apices as centers of moments and compute the values of M ; these divided by the depth of the truss give the strains, which are tensile if M is positive, compressive if M is negative. To find the lower chord strains, we choose the upper apices from which to measure the values of x, and divide the resulting values of M by the depth of the truss ; if M is positive these give compressive strains ; if negative, tensile ones.

Everything is thus known, except the shears and moments at the supports, and for these formulæ and methods will be presented in Chapter II, by which they may be found for all cases. We give here, however, two tables from which they may be found for the common case when all the spans are equal, and which, by a simple law, may be extended to include any number of such spans.

As before, let w be the uniformly distributed load per linear unit; let l be the length of each span, then will $w\,l$ be the weight of one span. The shear at any support is a fractional part of $w\,l$, or

$$\text{Shear}=\text{A}\ w\,l$$

A being a fraction given in the following triangle :

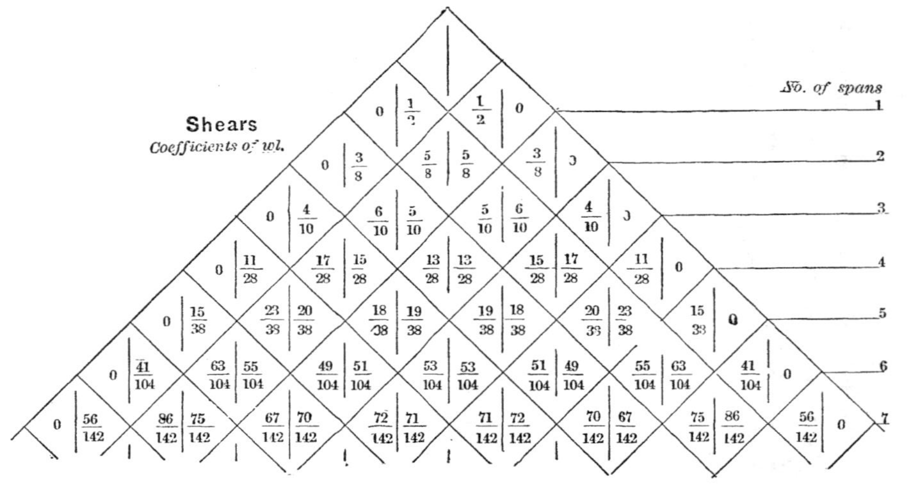
Shears
Coefficients of wl.
No. of spans
1
2
3
4
5
6
7

Each of the squares composing this triangle represents one of the supports, and its left hand division gives the left hand shear S'_{n-1}, and the right hand one the shear S_n (Fig. 3). Thus, in the girder of three spans, the triangle shows that the first support, beginning at the left, has on the left no shear, and on the right $\frac{4}{10}\,w\,l$, that the second support has on the left a shear of $\frac{6}{10}\,w\,l$, and on the right one of $\frac{5}{10}\,w\,l$. The sum of the two shears for any supports is of course its reaction. For example, a girder of six equal spans has at its middle support a reaction of $\frac{106}{104}\,w\,l$.

The moment at any support will be a fractional part of $w\,l$, or

$$\text{Moment} = \text{B}\,w\,l$$

B being a fraction given in the following triangle; in which like the preceding one the spaces indicate the supports of the girder. Thus, the fourth horizontal line refers to a girder of four spans, the moments at the first and last supports being 0, at the second and fourth $\frac{3}{28}\,w\,l^2$ and at the middle one $\frac{2}{28}\,w\,l^2$.

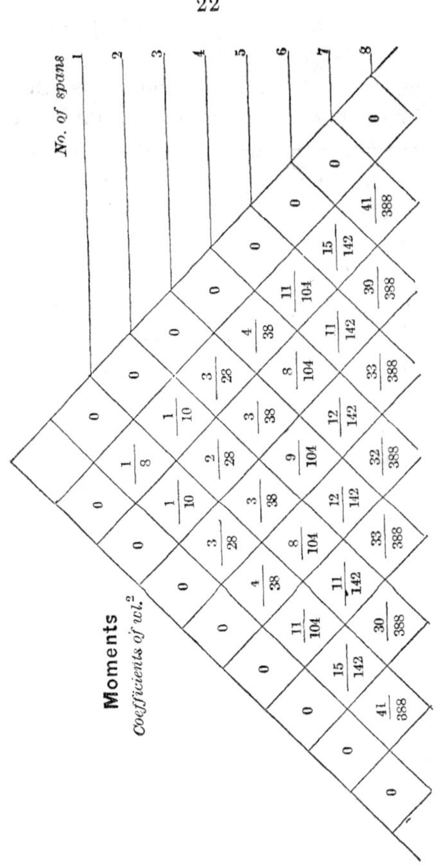
Moments
Coefficients of wl²
No. of spans
1
2
3
4
5
6
7
8
0 0
0 1/8 0
0 1/10 1/10 0
0 3/28 2/28 3/28 0
0 4/38 3/38 3/38 4/38 0
0 11/104 8/104 9/104 8/104 11/104 0
0 15/142 11/142 12/142 12/142 11/142 15/142 0
0 41/388 30/388 33/388 32/388 33/388 30/388 41/388 0

The triangles can be extended to any required length by the application of the following law which obtains in all *oblique* columns. Any fraction belonging to an *even* number of spans, may be obtained by multiplying both numerator and denominator of the preceding fraction by two and adding the numerator and denominator of the fraction preceding that. Thus for eight spans, the fraction $\frac{33}{388}$ is equal to $\frac{2\times 11+11}{2\times 142+104}$ or to $\frac{2\times 12+9}{2\times 142+104}$, according as we use one oblique column or the other. For an *odd* number of spans, any fraction is found by adding the two preceding fractions, numerator to numerator and denominator to denominator. Thus for seven spans,

$$\frac{12}{142}=\frac{8+4}{104+38} \text{ or } \frac{12}{142}=\frac{9+3}{104+38}.$$

These tables* furnish the data for solv-

* These triangles were first given by the author in the *Journal of the Franklin Institute* for March, 1875. A demonstration of the laws governing them may be seen in the same Journal for April, 1875.

ing all questions concerning continuous girders whose supports are on the same level, whose spans are all equal, and which are loaded uniformly throughout their entire length. The reader should first acquire facility in the use of the tables. We give, therefore, a few examples for practice :

1. In a girder of six spans, what is the reaction at the second support?

$$\textit{Ans.}\ R_2 = \frac{118}{104}\, w\, l.$$

2. In one of eight spans, what is the reaction at the middle support?

$$\textit{Ans.}\ R_5 = \frac{386}{388}\, w\, l.$$

3. In one of ten spans, what is the moment over the fourth support from the left?

$$\textit{Ans.}\ M_4 = \frac{123}{1448}\, w\, l^2.$$

4. In one of seven spans, what are the shears S_2 and S'_2? (see Fig. 3.)

$$\textit{Ans.}\ S_2 = \frac{75}{142}\, w\, l \quad S'_2 = \frac{67}{142}\, w\, l.$$

5. In one of six spans, what is the moment M_6 and the shear S_6?

$$Ans.\ M_6 = \frac{11}{104} w l^2 \quad S_6 = \frac{63}{104} w l^2.$$

Having thus found from the triangles, the moment M_n and the shear S_n for the nth support, the shear S and the moment M for any section in the nth span are readily found from the formulæ

$$S = S_n - w x$$
$$M = M_n - S_n x + \tfrac{1}{2} w x^2$$

which we have demonstrated above, and

Fig. 3.

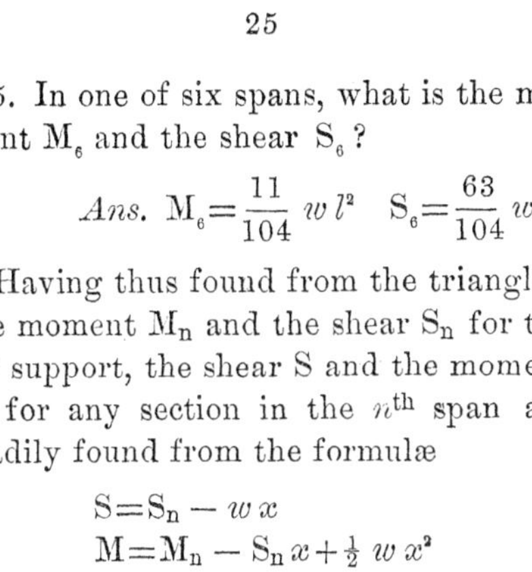

in which x is the distance from the support n to the assumed section. If in these x be made equal to l, they will, of course, give the shear S'_n at the left of the $n+1^{th}$ support, and the moment M_{n+1} over that support. We will illustrate their use by a few examples :

6. In a continuous girder of three spans, what is the shear and the moment at the center of the middle span?

We have from the table $S_2=\frac{5}{10}wl$ and $M_2=\frac{1}{10}wl^2$.

Hence

$$S=\frac{5}{10}wl-wx$$

$$M=\frac{1}{10}wl^2-\frac{5}{10}wlx+\tfrac{1}{2}wx^2$$

and placing x equal to $\frac{1}{2}l$, we have

$$S=0 \quad M=-\frac{1}{40}wl^2$$

7. In a girder of six spans, find the shear and moment at the center of the second span?

Ans. $S=\frac{3}{104}wl \quad M=-\frac{7}{208}wl^2$

8. In one of four spans, what is the shear and moment in the third span for $x=\frac{1}{4}l$ and $x=\frac{3}{4}l$?

Ans. $S=\frac{3}{14}wl$ and $S=-\frac{4}{14}wl$

$M=\frac{7}{224}wl^2$ and $M=\frac{1}{224}wl^2$.

In computing a framed truss we need to find the value of S for a section cutting every diagonal and that of M for one passing through each panel apex. from these we readily derive the strains in the webbing and chords by the rules explained above. A single example will render the whole process clear. (We here treat of the dead load only ; computations involving the live or rolling load will be presented hereafter.)

Let the truss represented in Fig. 2 consist of seven continuous spans, each sixty feet in length. Let the uniformly distributed load per linear foot be two hundred pounds, one half of which rests upon the upper chord and the other half upon the lower. The lower chord is divided into six bays, each of ten feet, and is connected with the upper one by a Warren system of diagonals. The depth of the truss is seven feet. Let it be required to compute the strains in all the pieces of the third span, due to this dead load.

We first take from the triangles for a

girder of seven spans, $S_3=\frac{70}{142}\,w\,l$ and $M_3=\frac{11}{142}\,w\,l^2$. We have for our case, $w=200$ lbs. and $l=60$ feet. Hence

$$S_3=5916 \text{ lbs. and } M_3=55775 \text{ lbs. ft.}$$

Inserting these and the value of w in the above general formulæ, we have

$$S=5916-w\,x$$
$$M=55775-5916\,x+100\,x^2$$

as the value of the shear and moment for any section x.

Now, since this is a framed truss, and the several pieces are to be subjected

Fig. 2.

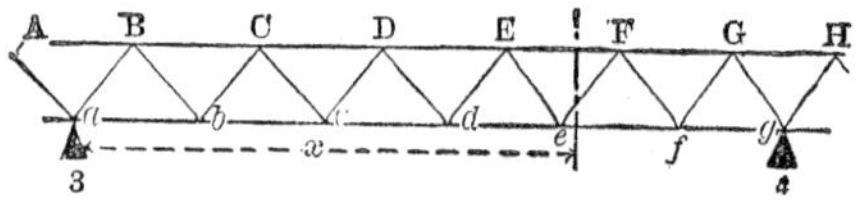

only to longitudinal strains, the load should not be strictly uniformly distributed but concentrated on the upper chord at the panel points B, C, etc. and on the lower chord at a, b, c, etc. Allowing that each of these points receives an equal weight and that a and g, count as but one point, we have at a 500 lbs., at g 500 lbs. and at each of the others 1000 lbs. In finding the shear for the diagonal a B, we pass the section anywhere between a and B and take wx as 500, for B b wx is 1500, for b C 2500 and so on; these subtracted from S_3 give the required shears. This is in fact nothing but taking the algebraic sum of all the exterior forces between the left hand of the truss and the diagonal under consideration, for S_3 is the sum of those forces from the left end to the beginning of the span. For the diagonal Fe we have, for example,

$$S = 5916 - 8500 = -2584 \text{ lbs.}$$

Thus by successive subtraction we find the shears for all diagonals. Multiplying them by the secant of the angle between a diagonal and the vertical or by

$$\text{Sec. } \theta = \frac{\sqrt{25+49}}{7} = 1.229$$

and we have the required strains. To determine their character we have simply to consider that *a positive shear causes a* $\left\{\begin{matrix}\textit{tensile}\\ \textit{compressive}\end{matrix}\right\}$ *strain in a diagonal which slopes* $\left\{\begin{matrix}\textit{upward}\\ \textit{downward}\end{matrix}\right\}$ *toward the left hand support*, while a negative shear produces the reverse. In the following table, the results thus determined are recapitulated. The first column shows the name of the diagonal corresponding to Fig. 2, the second gives the shears, the third gives the slopes, + indicating an upward inclination toward the left, and − a downward one, and the last column gives the strains, + indicating *tension*, and − *compression*. In forming the last column from the two preceding ones, it will be noticed that the rule of signs is observed :

DIAGONALS. (See Fig. 2.)

Piece.	Shear.	Slope.	Strain.
B *a*	+5416	−	−6656 lbs.
B *b*	+4416	+	+5427
C *b*	+3416	−	−4198
C *c*	+2416	+	+2969
D *c*	+1416	−	−1740
D *d*	+ 416	+	+ 511
E *d*	− 584	−	+ 718
E *e*	−1584	+	−1947
F *e*	−2584	−	+3176
F *f*	−3584	+	−4405
G *f*	−4584	−	+5634
G *g*	−5584	+	−6863

We will now pass to the computation of the chord strains. In the above expression for M the quantity $\frac{1}{2} w x^2$ or $100 x^2$ is the moment of the load between the point 3 and the assumed section and its value is the same whether the load be considered as uniformly distributed or concentrated at the apices as above. Hence to find the moments for the upper chord we have in the expression

$$M = 55775 - 5916 x + 100 x^2$$

simply to give to x the successive values

0, 10, 20, etc., since for the bays A B, B C, C D, etc., the opposite vertices *a*, *b*, *c* etc., must be taken as centers of moments. Thus if the bay C D be cut rotation will at once begin around the point *c*; we take then $x = 20$ and find for the moment of the strain in C D,

$$M = -22545 \text{ lbs. ft.}$$

and dividing this by its lever arm or seven feet we have 3221 lbs for the strain. The character of the strain is found by recollecting that *a positive moment causes a* $\left\{\begin{matrix} \textit{tensile} \\ \textit{compressive} \end{matrix}\right\}$ *strain in the* $\left\{\begin{matrix} \textit{upper} \\ \textit{lower} \end{matrix}\right\}$ *chord*, while a negative moment produces the reverse. If we designate then tension by + and compression by —, the signs of the strains in the upper chord will be the same as those of the moments. In this way it is easy to compute the following results :

Upper Chord. (See Fig. 2).

Bay.	Moment.	Strain.
A B	+55775 lbs. ft.	+7968 lbs.
B C	+ 6615	+ 945
C D	−22545	−3221
D E	−31705	−4529
E F	−20865	−2981
F G	+ 9985	+1426
G H	+60845	+8692

For the lower chord the calculation is very similar. The centres of moments are taken at the points B, C, etc., the successive values of x are 5, 15, 25, etc., the strains are numerically one-seventh of the moments, and their signs are opposite to that of the moments. Thus for the bay *ef*, $x=45$, $M= -7945$ lbs. ft. and the strain in *ef* is 1135 lbs. tension. The results are in the following table :

LOWER CHORD. (See Fig. 2.)

Bay.	Moment.	Strain.
ab	+28695 lbs. ft.	−4099 lbs.
bc	−10465	+1495
cd	−29625	+4218
de	−28785	+4112
ef	− 7945	+1135
fg	+32895	−4699

and the strain sheet for the span is now complete.

In the same way the strains for each of the other spans may be readily found. From the symmetry of the truss it is evident that the fifth span will be exactly the same as the third, the sixth the same as the second, and the seventh the same as the first. For the fourth span the value of S and M for any section x are

$$S=6000-w\,x$$

$$M=60848-6000x+100\,x^2$$

and the strains will be the same on each side of its center. For the first span S_1 is the same as the reaction R_1 the moment M_1 is zero and we have

$$S = 4732 - w\,x$$
$$M = -4732\,x + 100\,x^2$$

It will be seen then that the computation of the strains in a continuous girder is exactly the same as in a simple one, except only in the preliminary determination of the shears and moments at the supports. In a simple girder the end shears are the same as the reactions, which are known from the law of the lever, and the moments at the supports are zero. In a continuous one these quantities must be determined by formulæ, or, for the case of equal spans uniformly loaded,* taken from the triangles which we have given above. They may also be found by a graphical process.

If the truss above discussed were built with seven simple girders, the strains in each would be the same. It may prove interesting then to compare the results above found with those for a simple gir-

* Other convenient tables for concentrated loads and for uniform loads over one span only are given in the article above referred to in the *Journal of the Franklin Institute.*

der. The mode of computation is essentially the same ; the end shears are each one half the total load, the pieces A B and G H in Fig. 2 disappear, and for any section x we have

$$S=6000-wx$$
$$M=-6000x+100x^2$$

Considering as before that the load is concentrated at the panel points we have 500 lbs. at a and g and 1000 lbs. at B, C, D, b, c, d, etc., respectively. We then find the shears and moments and from them deduce the strains as above described. The results are given below compared with those for the third span of the continuous truss.

(*See Table on following pages.*)

Adding these strains regardless of sign we find the two sums to be the same. It can be easily demonstrated that for the dead load such should be

DIAGONALS. (See Fig. 2.)

Piece.	Continuous truss.	Simple truss.
B *a*	—6656 lbs.	—6760 lbs.
B *b*	+5427	+5531
C *b*	—4198	—4301
C *c*	+2969	+3072
D *c*	—1740	—1844
D *d*	+ 511	+ 614
E *d*	+ 718	+ 614
E *e*	—1947	—1844
F *e*	+3176	+3072
F *f*	—4405	—4301
G *f*	+5634	+5531
G *g*	—6863	—6760
Sums....	44244 lbs.	44244 lbs.

the case. As far as the diagonals are concerned, the two structures require an equal amount of material.

UPPER CHORD. (See Fig. 2.)

Bay.	Continuous truss.	Simple truss.
A B	+7968 lbs.	
B C	+ 945	— 7143 lbs.
C D	—3221	—11429
D E	—4529	—12857
E F	—2981	—11429
F G	+1426	— 7143
G H	+8692	
Sums....	29762 lbs.	50001 lbs.

Adding the strains in the upper chord we observe that the sum for the simple truss is about 1.7 times that of the other. If the amount of material is to be proportional to the strain, a considerable saving will here be expected.

Lower Chord. (See Fig. 2.)

Bay.	Continuous truss.	Simple truss.
ab	—4099 lbs.	— 3929 lbs.
bc	+1495	— 9643
cd	+4218	—12500
de	+4112	—12500
ef	+1135	— 9643
fg	—4699	— 3929
Sum....	19758 lbs.	52144 lbs.

The lower chord in the simple truss would then be subjected to about 2.6 times as much strain as in the continuous one.

If we suppose that the same working strength may be allowed for compression as for tension, we may obtain an estimate of the saving in material by employing a continuous truss instead of a simple one. The amount of material will be proportional to the strain and to the length of the piece strained. Regarding the bays of the chord as unity, the diagonals will be represented in length by 0.86. The proportionate amounts of iron will then be found by multiplying the above sums by unity for the chords and by 0.86 for the diagonals. Thus we have a

COMPARISON.

	Continuous truss.	Simple truss.
Diagonals	38050	38050
Upper Chord..	29762	50001
Lower Chord..	19758	52144
Total.........	87570	140195

from which we see that the amounts of material in the two cases are in the ratio of the numbers 87570 and 140195 or as 1 to 1.6. For this particular span then a saving in material of thirty-seven and a half per cent. is effected by using a continuous truss instead of a common one.

It is capable of demonstration that for girders subjected only to dead load, the total amount of strain in the webbing will be the same for simple as for continuous trusses, and also that under the most favorable circumstances, the total strain in the chords of the first is to that in the chords of the second as $\sqrt{27}$ is to 2 or nearly as 2.6 to 1.

In studying the theory of girders many interesting questions arise which are of little importance in practice. One of these is the determination of the inflection points. At these points the curvature of the beam changes, the strain passes from tension to compression and the moment is zero. At any point in the nth span the moment is

$$M = M_n - S_n x + \frac{1}{2} w x^2$$

Making in this M equal to zero and solving the equation with reference to x we find

$$x = \frac{S_n}{w} \pm \sqrt{\frac{S_n^2}{w^2} - \frac{2 M_n}{w}}$$

For a girder of equal spans and uniformly loaded we may hence write for the two inflection points,

$$x = A\,l \pm l\sqrt{A^2 - 2B}$$

in which A and B are to be taken from the above triangles, A always being taken for the right hand side of the support under consideration, for example, in a girder of eight spans the inflection points for the fourth span are at the points

$$x = \frac{195}{388}\,l \pm l\sqrt{\left(\frac{195}{388}\right)^2 - \frac{66}{388}}$$

or for $x = 0.22\,l$ and $x = 0.79\,l$.

The point of maximum moment, or the point near the center of the beam, where the chord strain is the greatest is more important and readily determined from the above general value for M. Differentiating it with reference to x we have

$$\frac{dM}{dx} = -S_n + w\,x = 0$$

that is, *the maximum moment obtains at*

the point where the shear is zero. Its value is found by replacing x by $\frac{S_n}{w}$ which gives

$$\text{Max. } M = M_n - \frac{S^2_n}{2w}$$

as the greatest negative moment.

The following examples will enable the reader to test his knowledge of the preceding principles :

9. In a girder of two spans uniformly loaded what are the maximum positive and negative moments ?

Ans. $0.125\ w l^2$ and $-0.071\ w l^2$.

10. In one of three spans what is the maximum negative moment in the middle span ?

Ans. $-0.025\ w l^2$.

11. In one of eight spans where are the inflection points in the fifth span ?

Ans. $x = 0.21\ l$ and $x = 0.78\ l$.

12. A continuous girder of three spans, each equal to fifty feet, is divided into

five panels on the lower chord, and has bracing similar to that shown in Fig. 2. Supposing a load of five tons applied at each of the lower panel points, what are the strains in each of the pieces of the middle span? the height of the truss being six feet.

Ans. In $a\,b$, -11.7 tons; in $b\,c$, $+1.7$; in $B\,b$, $+13$; etc.

In this chapter we have treated of the continuous girder when affected only by dead load or its own weight. In the following chapters we shall take up the action of the live or variable load.

CHAPTER II.

A continuous girder, loaded in any manner, is held in equilibrium by the upward pressures or reactions of the supports, and, as we have seen, these reactions are alone sufficient for the complete determination of the strains in every part of the girder. But if, regarding the question as one of pure statics, we consider the beam as rigid, we find it impossible to determine the reactions when the number of supports is greater than two. This does not arise from the fact that in an actual case the question is indeterminate, but simply because in considering the girder as rigid we have restricted the data to the mere weight, neglecting entirely the physical properties of the material. By taking into account the *elas-*

ticity of the girder, the problem becomes determinate; we find the reactions, or what is equivalent, the shears and moments at the supports, and from these the investigation of the internal forces or strains is easy.

THE ELASTIC LINE.

When a girder is acted upon by vertical forces, a change of shape arises, which causes the originally parallel fibers, to be on one side lengthened, and on the other shortened. Between the lengthened and shortened fibers, there is a plane which undergoes no change of length; the central line of this plane is called the *neutral axis* or the *elastic line.* Thus, in the bent beam represented in Fig. 4, $m\ o$ is the neutral axis, the fibers above it being shortened or compressed, and those below it lengthened or tensioned.

We derive the equation of the elastic line upon three hypotheses: 1*st*, that all planes perpendicular to the neutral axis before the bending or flexure, preserve during the bending their perpendicularity and their form as planes ; 2*d*, that the change of length of a body subjected to a force is, within certain limits called the elastic limits, proportional to the intensity of the force ; and 3*d*, that the change of shape is so little that the length of the neutral axis is sensibly the same as its horizontal projection.

In Fig. 4 we have a longitudinal section of a portion of a bent beam; the two planes $a\,b$ and $d\,e$, originally parallel, remaining perpendicular to the neutral axis $m\,o$, and intersecting in c the center of curvature. Hence, drawing $f\,g$ parallel to $a\,b$ through o, the lines fd, ge, etc., denote the elongations and compressions of the respective fibers, and we sec from the figure that

$$o\,d : o\,d' :: d\,f : d'\,f'$$

or the change of length in the fibers is proportional to their distances from the neutral axis. This is the consequence of the first hypothesis.

Designating by H and H′ the force

acting in the fibers df and $d'f'$, the second hypothesis says that

$$H : H' :: df : d'f'$$

Fig. 4.

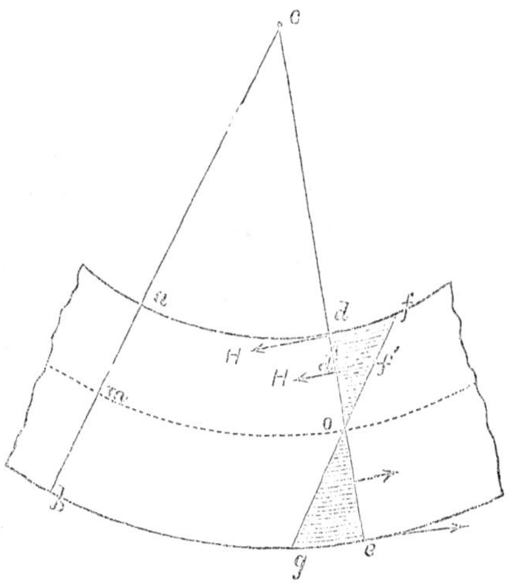

Hence, by combining these two proportions,

$$H : H' :: od : od'$$

or, the horizontal forces are directly proportional to their distances from the neutral axis. Therefore, if we denote the distance of *any* fiber from the axis by z the strain upon it by H′, the distance of the remotest fiber by e and the strain upon it by H, we have

$$H' : H :: z : e \quad \text{or } H' = \frac{H\,z}{e}$$

We have thus far considered the cross-section of the fibers as unity. If the actual section be a, the force in each is evidently $\frac{H\,a\,z}{e}$. Each of these forces, as for instance H′ in the figure, tends to turn the beam around the point o with a lever arm $o\,d'$ or z, and its moment or the measure of that tendency to rotation is the product of the force $\frac{H\,a\,z}{e}$ by the distance z, or $\frac{H\,a\,z^2}{e}$. The sum of all these moments is

$$M = \frac{H}{e}\,\Sigma\,a\,z^2$$

or, since $\Sigma a z^2$ is the moment of inertia of the section $a\,b$, we have

$$M=\frac{H\,I}{e}$$

as the expression for the sum of the moments of the internal forces, H being the strain in the remotest fiber, e its distance from the neutral axis, and I the moment of inertia of the cross-section.

The line $d f$ denotes the change of length of the fiber $a\,d$ due to the force H. Hence if E be the coefficient of elasticity,*

$$a\,d : d f :: E : H$$

Designating the radius $c\,o$ by r we have from the similar figures $o\,d f$ and $c\,a\,d$ (since $m\,o=a\,d$),

$$a\,d : d f :: r : e$$

* The *Coefficient of Elasticity* is the ratio of the force of displacement to the amount of displacement taken upon a cube whose edge is unity; Hence for the above case $E=H\div\frac{d f}{ad}$. The term modulus of elasticity properly relates to the impact of bodies, and is a measure of elasticity in the common sense of the word, unity indicating perfect elasticity or restitution of form. These terms are often confounded by writers.

Combining these proportions we find

$$\frac{H}{e}=\frac{E}{r}$$

and hence, for the value of the internal moment, we have

$$M=\frac{E\,I}{r}$$

The radius of curvature of any plane curve, whose length is u, and co-ordinates x and y is*

$$r=\frac{d\,u^3}{d\,x\,d^2\,y}$$

And as by our third hypothesis we may place $d\,u=d\,x$, this becomes

$$r=\frac{d\,x^2}{d^2\,y}$$

which, inserted in the above value of M, gives

$$(1)\qquad \frac{d^2\,y}{d\,x^2}=\frac{M}{E\,I}$$

as the differential equation of the elastic

---

* See any work on the Differential Calculus.

curve, applicable to all bodies subject to flexure, which fulfill the condition imposed by the third hypothesis. The coefficient of elasticity and the moment of inertia may be different in every section.

CONDITIONS OF EQUILIBRIUM.

Let us consider the rth span of a continuous girder whose length is l_r, and let a single concentrated load P_r be placed on this span at a distance kl_r from the left hand support r. This load, the loads on the other spans, and the weight of the girder itself, are held in equilibrium by the vertical reactions R_{r-1}, R_r, etc., of the several supports. (See Fig. 5.)

Let us pass a section between the load P_r and the support $r+1$ at a distance x from the rth support. As shown in the last chapter, all the internal forces in this section are represented by a shear S and a moment M. The shear S is equal to the algebraic sum of all the external forces upon the left hand side of the section, and the moment M is equal to the sum of the moments of those forces with respect to the section as a center. Hence, regarding upward forces as positive, and a moment as positive when it tends to cause tension in the upper fiber of the section, we have

$$(2) \qquad \begin{aligned} S &= S_r - P_r \\ M &= M_r - S_r x + P_r (x - kl_r \end{aligned}$$

in which S_r is the shear at the right of the rth support, and M_r the moment at that support. In like manner for a section between r and P_r, we have

$$\begin{aligned} S &= S_r \\ M &= M_r - S_r x \end{aligned}$$

The internal forces at any section can then be found as soon as the shear and moment at the preceding support are known.

If, in the above expression, we make x equal to l_r, M becomes M_{r+1}, and we deduce

$$(3) \quad S_r = \frac{M_r - M_{r+1}}{l_r} + P_r (1-k)$$

The shear and moment at any section can then be determined as soon as M_r and M_{r+1}, the moments at the preceding and following supports, are known.

These conditions of equilibrium are entirely independent of variations in the dimensions or material of the beam, or in the relative heights of the supports of the girder.

THE EQUATION OF THE ELASTIC LINE.

In order to apply equation (1) to the case of continuous girders, we have to insert for M, E and I their values as functions of x and integrate the equation twice. E, however, cannot under any ordinary circumstances be a function of x, it being dependent upon the elasticity of the material alone, which is nearly the same in one and the same beam, and we hence regard it as constant. In a beam of uniform section I is constant, and although it varies in common bridge trusses, we shall be obliged in order to bring the investigation within the limits of this paper to consider it always as constant, taking care to point out afterwards the slight error thus introduced. Inserting then in (1) the value of M from (2), we have

$$\frac{d^2 y}{d x^2}=\frac{M_r - S_r x + P_r (x - k l_r)}{E I}$$

as the differential equation, applicable to girders of constant elasticity and uniform cross section. Integrating this, the constant is t_r, the tangent of the inclination of the elastic line at the support r and

$$(4) \quad \frac{d y}{d x}=t_r+\frac{2 M_r x - S_r x^2 + P_r (x - k l_r)^2}{2 E I}$$

we have thus far taken no account of the relative heights of the supports. For the reasons mentioned in the last chapter, we shall consider them as all upon the same level. The constant for the second integration is then 0, the origin being at r, and we have

$$(5) \quad y=t_r x+\frac{3 M_r x^2 - S_r x^3 + P_r (x - k l_r)^3}{6 E I}$$

as the equation of the elastic curve between the load P_r and $r+1^{th}$ support (Fig 5). If there be several loads we have only to affix the sign of summation Σ to the term involving P_r, and if that term be omitted we shall have the equation between the load and the rth support, since for any section between those points $M=M_r-S_r x$.

If in (5) we make $x=l_r$, y becomes 0, and inserting for S_r its value from (3), we find

$$(6)\quad 6\,E\,I\,t_r=-2\,M_r\,l_r-M_{r+1}\,l_r+P_r\,l_r^2\,(2\,k-3\,k^2+k^3)$$

Thus the equation of the curve is completely determined, when we know M_r and M_{r+1} the moments at the supports r and $r+1$. These may be found by the remarkable theorem of three moments.

THE THEOREM OF THREE MOMENTS.

In Fig. 5 is represented a portion of a continuous truss. Beginning at the left hand end, the lengths of the spans are denoted by l_1, l_2, ... l_r, etc, and the supports are designated as 1, 2, ... r, etc. Upon the spans l_{r-1} and l_r are loads P_{r-1} and P_r, whose distances from the nearest left hand supports are $k\,l_{r-1}$ and $k\,l_r$, k being any fraction less than unity, and not necessarily the same in the two cases. The equation of the elastic line between P_r and the support $r+1$ is given by (5), and the tangent of the angle which the curve at the section x makes with the axis of abscissæ is given by (4).

FIG. 5.

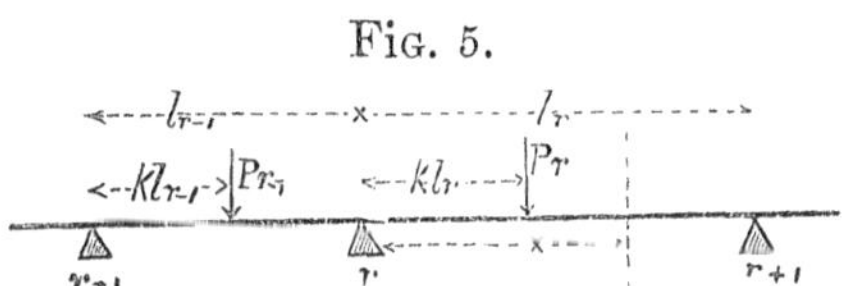

If in (4) we substitute for S_r its value from (3), and for t_r its value from (6), and make $x=l_r$, $\frac{dy}{dx}$ becomes t_{r+1} the tangent of the inclination of the curve at $r+1$, and we find

$$6\,EI\,t_{r+1}=M_r\,l_r+2\,M_{r+1}\,l_r-P_r\,l_r^2(k-k^3)$$

Now, if we consider the origin moved from the support r back to $r-1$, we may derive a value for t_r by simply diminishing each of the indices in the above expression by unity, hence

$$(7)\quad 6\,E\,I\,t_r = M_{r-1}\,l_{r-1} + 2\,M_r\,l_{r-1} \\ - P_{r-1}\,l^2_{r-1}\,(k-k^3)$$

Equating the values of $6\,E\,I\,t_r$ given by (6) and (7), we have

$$(8)\quad M_{r-1}\,l_{r-1} + 2\,M_r(l_{r-1}+l_r) + M_{r+1}\,l^r \\ = P_{r-1}\,l^2_{r-1}\,(k-k^3) + P_r\,l_r^2(2\,k-3k^2+k^3)$$

which is the most general form of the theorem of Three Moments for girders of constant cross-section. By prefixing the sign Σ to the terms in the second member, it becomes applicable to any number of single loads. For uniform loads w_{r-1} and w_r per linear unit, we have only to place

$$P_{r-1} = w_{r-1}\,d\,(k\,l_{r-1}) = w_{r-1}\,l_{r-1}\,d\,k$$
$$P_r = w_r\,d\,(k\,l_r) = w_r\,l_r\,d\,k$$

And to replace the sign of summation Σ by that of integration $\int$. If these loads extend over the entire spans l_r and l_{r-1}, we take the integrals between the limits $k=0$ and $k=1$, and have

$$M_{r-1}\, l_{r-1} + 2\, M_r\, (l_r + l_{r-1}) + M_{r+1}\, l_r = \frac{w_{r-1}\, l^3_{r-1}}{4} + \frac{w_r\, l^3_r}{4}$$

which is the theorem as first announced by Clapeyron.

For each support of a continuous girder an equation may be therefore written involving the moment at that support, and those at the preceding and following support. In a girder of s spans there are $s+1$ supports, and since the moments at the first and last supports are zero we have $s-1$ moments, whose values may be found by the solution of the $s-1$ equations. The moments give the shear at any support, and by (2) the internal forces or strains may be determined for every section of the girder.

REMARKS ON THE PRECEDING THEORY.

The laws of the theory of continuity above deduced can be regarded as only approximate. Of the three hypotheses upon which the differential equation of the elastic line is deduced, the first, although a most reasonable assumption, has not been definitely verified by experiment, and the second is rendered somewhat doubtful by the extreme difficulty in delicate experiments of assigning the elastic limits. Nevertheless they are universally regarded by all writers as sufficiently accurate to form the basis of a working theory, and must continue to be thus used until we attain to a more thorough knowledge of matter and force. The third hypothesis, however, is a limitation of the data, which we are at perfect liberty to make, since we know that the increase in length of the girder by deflection is too small to be practically measured. We may conclude then that the equation (1) is an extremely close approximation to the actual law governing straight elastic beams. From the time of Navier to the present it has been so accepted and used.

The next hypothesis or limitation of

data, which we make, is that E, the coefficient of elasticity, is constant throughout the girder. In a solid beam of ordinary homogeneous material, there can be no reason for supposing it otherwise.

In the *Journal of the American Society of Civil Engineers* for May, 1876, appeared an article by Charles Bender, C. E., in which the use of continuous bridges is strongly opposed. One of his main objections is—that the theory upon which such bridges are computed is unreliable in consequence of the assumption of a constant coefficient of elasticity, and he quotes the records of experimenters to show that values for the coefficient of iron and steel have been observed ranging from 17,000,000 to 40,000,000 pounds per square inch. These limiting values are, however, decidedly exceptional, but even granting that such variations may exist in materials and forms like soft iron wire, steel rails and wrought iron eyebars, it cannot be supposed that they will occur in one and the same structure, where the material is of one kind, of similar cross sections and which has been subjected in the same mill to the same process of manufacture. The mere statement that Morin has observed values of the coefficient of elasticity as low as 17,000,000 has very little weight when unaccompanied by any reference to the kind of iron experimented upon. Let us see what Morin himself actually says in recapitulating the results of experiments upon wrought iron.*

" Iron of *superior quality*, which comes from standard ores, and which has been manufactured exclusively with charcoal, or iron from sheet metal, many times refined, may give for the coefficient of elasticity values as high as E=28,400,000, or even E=31,200,000 lbs. per square inch, equal to those furnished by ordinary steel. Iron of *ordinary* manufacture reduced with common coal, and drawn into forms like rails, T irons,

*Morin; *Resistance des Materiaux*, Paris, 1862. Vol. I, p. 443.

flanges, etc., give such values as E = 24,100,000 and E = 25,600,000. Finally the *most soft and ductile* iron furnishes values as low as E=21,300,000, or E=19,800,000, or even E=17,000,000. It is well, then, in calculations upon the strength of iron, to ascertain the quality of the material and the process of manufacture."

Mr. Bender likewise alludes to experiments upon wrought iron bars in which the coefficient of elasticity was found to be 40,000,000, and it seems to be implied by his language that such values are of common occurrence. The fact, however, that a standard writer on the strength of materials, like Morin, regards 34,000,000 as an exceptionally high figure for iron, may justify us in demanding that when a value like 40,000,000 is quoted, we should be furnished with some details concerning the quality of such iron, the process of manufacture, as well as a description of the testing machine and the manner of measuring the small extensions or compressions, from which the coefficient is calculated, or at least that we should be referred to the book or journal where such experiments are described. And further as it is well known that by straining a bar beyond the elastic limits, very low values of E can be deduced, are we not justified in asking similar information concerning experiments which furnish such values?

Undoubtedly there is some variation in the elasticity of different pieces of iron, even when great care has been taken to ensure uniformity of material and manufacture, and it is greatly to be desired that experiments should be made to determine how it varies with the cross section and length of the piece. As soon as such a law of variation is discovered (if any exist), we shall be obliged to consider E as variable in investigating a continuous truss. But if no law exists and we know only the fact that there are slight variations in the elasticity of different pieces in the same truss, what is to be done? Nothing but to regard E as constant, being well assured that the distribution of the variable pieces throughout the structure will be governed by the law of probability, and that hence the girder as a whole will conform closely to the theoretic form of the elastic line.

The next argument which it is our duty to criticise in that article is, that the theory of continuous girders is unreliable, because the calculated deflection does not generally agree with the actually measured deflection. The accuracy of the computed strains must depend upon the accuracy of the theorem of three moments, and this it is asserted depends upon the calculated deflection. And because the deflection as actually measured is sometimes no more than one-half of the calculated one, hence, it is said, the same differences may occur in the strains, and the whole theory is unworthy of consideration.

This we can only regard as a striking instance of the incompetency of practical men to draw conclusions from even simple experiments. The reader who has followed onr presentation of the theory of the elastic line, will see at once that the value of the deflection given by (5) only enters the discussion as an auxiliary for finding (6), the tangent of the inclination angle at the support. The process supposes, indeed, that E is constant, but it supposes nothing whatever concerning the value of the deflection at any point. When we pass to the next span and find again in (7) a second value of the tangent, the actual value of the deflection there is likewise not considered. And when by the combination of (6) and (7), we deduce (8) in which E does not appear, its very absence is a proof that the moments and hence the strains are entirely independent of its value or of the actual deflection. If two trusses of the same spans, height and form are continuous over several supports, one of steel having a coefficient of elasticity of 31,000,000 lbs. per square inch, and the other of wood having a coefficient of only one-twentieth as much, the reactions, shears, moments and strains would be the same in each. The measurement of the actual deflections of these bridges under given loads is only useful for determining and comparing their stiffness or elasticity, or in connection with theory for finding the values of E and I. The theory of continuity rests not upon absolute deflections, but on relative ones—on the *form* of the elastic curve, and this again upon the three universally accepted hypotheses included in our equation (1), with the additional assumption that the coefficient of elasticity is practically constant.

One more remark and we close for to-day a discussion which shall be resumed in our next chapter. Mr. Bender advises us to abandon the theory of flexure, to make no further advance in bridge building, to remain content with the simple lever, or at the utmost with the continuous (*sic*) patent hinged truss. But until such advice is enforced by more logical arguments than we have yet seen, we must continue our work in support of a theory and system which is universally accepted as only slightly deviating from the exact existing conditions, which is applied in the erection of continuous bridges by every nation except our own, and which perhaps if carried out by us might lead to more perfect structures than the world has yet seen. The great majority of coefficients of elasticity quoted in his paper, made by such men as Staudinger, Baker, Morin and Wöhler, were in fact *found by measuring the deflections of beams, and then from the theory of flexure computing the value of* E. He accepts those values, and on their evidence condemns the theory by which they were deduced! Is not Morin's conclusion, which we have quoted above, by far the most reasonable?

MOMENTS AT THE SUPPORTS.

The theorem of three moments given by (8) furnishes the means of finding the moments at the supports due to any assigned system of loading. The actual solution of those equations is, however, quite tedious when the number of spans is large, and we proceed therefore to develop a general solution, by which the values of the moments may be formulated and placed in a convenient form for numerical computation.

In the designing of continuous bridges it is only necessary to consider single loads concentrated at the panel apices or uniform loads extending over an entire span. Let us, then, consider a continuous girder of constant cross-section and homogeneous material whose supports are on the same level. Let, as in Fig. 6, the supports beginning at the left hand end be designated by the indices 1, 2, 3, ... r, etc, and the lengths of the spans l_1, l_2, l_3, ... l_r, etc. Call the number of spans s; then the last span will be l_s, and the last support $s+1$. The ends of the girders rest upon abutments in the usual manner, the lengths of the spans may be all different and their number may range from one to infinity. In the span l_r let a single load P be placed at a distance $k\,l_r$ from the support r, (k being any fraction less than

FIG. 6.

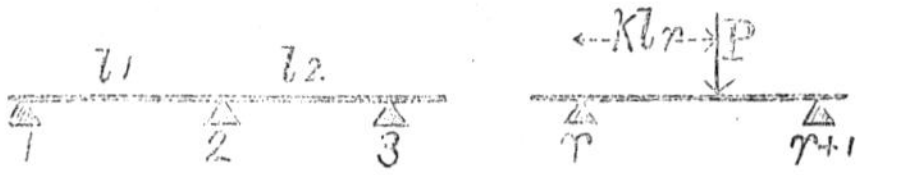

unity), or let this span be loaded uniformly from r to $r+1$ with a weight W, (W being equal to $w\,l_r$, if w is the load per linear unit), all the other spans being unloaded. By reference to (8) we notice that there are two functions of such loads which enter the equations of moments. If the single load P is alone considered these functions are

$$P\,l_r^2\,(2\,k-3\,k^2+k^3)$$
$$P\,l_r^2\,(k-k^3)$$

the first entering into the equation for the preceding support r, and the second into the one for the following support $r+1$. If the uniform load over the whole span is alone considered, these become each equal to $\frac{1}{4}\,W\,l_r^2$, as we have shown above in discussing the theorem of three moments. In the following investigation, we place therefore for abbreviation

$$\left.\begin{array}{l} A=P\,l_r^2\,(2\,k-3\,k^2+k^3) \\ B=P\,l_r^2\,(k-k^3) \end{array}\right\}\ \text{for a single load in span } l_r.$$

(I)

$$\left.\begin{array}{l} A=\tfrac{1}{4}\,W\,l_r^2 \\ B=\tfrac{1}{4}\,W\,l_r^2 \end{array}\right\}\ \text{for a uniform load over whole span } l_r.$$

By introducing the letters A and B to represent these functions, our discussion will apply equally well to a single load P, or to a weight W uniformly distributed over the whole of a single span.

Since the girder is not fastened at the abutments 1 and $s+1$, the moments at those points will be zero. The moments at 2, 3, ... r, etc., we designate by M_2, M_3, ... M_r, etc., and from (8) we may write an equation for each of those supports. As there is no load considered except on the span l_r, the right hand member of the equation for the support r will be A, of that for $r+1$ will be B and of all the others will be zero. Thus we have the following equations :

$$2\,M_2\,(l_1+l_2)+M_3\,l_2=0$$
$$M_2\,l_2+2\,M_3\,(l_2+l_3)+M_4\,l_3=0$$
$$\cdots\cdots\cdots\cdots\cdots$$
$$M_{r-1}\,l_{r-1}+2\,M_r\,(l_{r-1}+l_r)+M_{r+1}\,l_r=A$$
$$M_r\,l_r+2\,M_{r+1}\,(l_r+l_{r+1})+M_{r+2}\,l_{r+1}=B$$
$$\cdots\cdots\cdots\cdots\cdots$$
$$M_{s-2}\,l_{s-2}+2\,M_{s-1}(l_{s-2}+l_{s-1}) + M_s\,l_{s-1}=0$$
$$M_{s-1}\,l_{s-1}+2\,M_s\,(l_{s-1}+l_s) \qquad =0$$

The number of these equations is $s-1$, the same as the number of unknown moments. Their solution is best effected by the method of indeterminate multipliers. Let then the first equation be multiplied by a number c_2, the second by c_3, etc., the index of the as yet indeterminate numbers corresponding with that of the M in the middle term. Then let all the equations, thus multiplied, be added, and the resulting equation be arranged according to the coefficients of the unknown moments M_2, M_3, etc. Now, if we require that such relations exist between the multipliers, that all the terms in the first member shall reduce to zero, except the last containing M_s, the value of M_s is

$$M_s = \frac{A\,c_r + B\,c_{r+1}}{c_{s-1}\,l_{s-1} + 2\,c_s\,(l_{s-1} + l_s)}$$

And the values of the multipliers will be given by the equations

$$2\,c_2\,(l_1 + l_2) + c_3\,l_2 = 0$$
$$c_2\,l_2 + 2\,c_3\,(l_2 + l_3) + c_4\,l_3 = 0$$
$$c_3\,l_3 + 2\,c_4\,(l_3 + l_4) + c_5\,l_4 = 0$$
$$\text{etc.,} \qquad \text{etc.} \qquad \ldots\,.$$

After deducing the values of c from these equations, the value M_s is at once known.

Again, if we multiply the equations of moments, beginning with the last, by the indeterminate numbers d_2, d_3, etc., all the moments except M_2 may be eliminated, and we have

$$M_2 = \frac{A\,d_{s-r+2} + B\,d_{s-r+1}}{d_{s-1}\,l_2 + 2\,d_s\,(l_1 + l_2)}$$

and the multipliers will be given by the equations

$$2\,d_2\,(l_s + l_{s-1}) + d_3\,l_{s-1} = 0$$
$$d_3\,l_{s-1} + 2\,d_3\,(l_{s-1} + l_{s-2}) + d_4\,l_{s-2} = 0$$
$$\text{etc.,} \qquad \text{etc.,} \qquad \text{etc.}$$

The values of the numbers in the series c and d need only satisfy the equations as given above. Assuming then $c_2=1$, and $d_2=1$, we get the following :

$$
\begin{aligned}
&c_1=0\\
&c_2=1\\
&c_3=-2\frac{l_1+l_2}{l_2}\\
\text{(II)}\qquad &c_4=-2\,c_3-(2\,c_3+c_2)\,\frac{l_2}{l_3}\\
&c_5=-2\,c_4-(2\,c_4+c_3)\,\frac{l_3}{l_4}\\
&c_6=+2\,c_5-(2\,c_5+c_4)\,\frac{l_4}{l_5}\\
&\text{etc.,}\qquad\qquad \text{etc.}
\end{aligned}
$$

$$
\begin{aligned}
&d_1=0\\
&d_2=1\\
&d_3=-2\,\frac{l_s+l_{s-1}}{l_{s-1}}\\
&d_4=-2\,d_3-(2\,d_3+d_2)\,\frac{l_{s-1}}{l_{s-2}}\\
&d_5=-2\,d_4-(2\,d_4+d_3)\,\frac{l_{s-2}}{l_{s-3}}\\
&d_6=-2\,d_5-(2\,d_5+d_4)\,\frac{l_{s-3}}{l_{s-4}}\\
&\text{etc.,}\qquad\qquad \text{etc.}
\end{aligned}
$$

which reduce to numerical form as soon as the lengths of the spans for any particular case are substituted.

Since the $s-1$ equations of moments are of the same form as the equations of the multipliers c and d, we must have

$$M_3=c_3 M_2, \quad M_4=c_4 M_2, \text{ etc.,}$$
$$M_{s-1}=d_3 M_s, \quad M_{s-2}=d_4 M_s, \text{ etc.,}$$

or, if n indicate the index of any support,

$$M_n = c_n M_2 \quad \text{when } n<r+1$$
$$M_n = d_{s-n+2} M_s, \text{ when } n>r$$

Inserting in these the values of M_2 and M_s as found above, we have

$$M_n = c_n \frac{A\, d_{s-r+2} + B\, d_{s-r+1}}{d_{s-1}\, l_2 + 2\, d_s\, (l_1 + l_2)},$$

(III) when $n<r+1$

$$M_n = d_{s-n+2} \frac{A\, c_r + B\, c_{r+1}}{c_{s-1}\, l_{s-1} + 2\, c_s\, (l_{s-1} + l_s)},$$

when $n>r$

which give the values of the moments at all supports in terms of the quantities A and B, depending only upon the character of the load and its position in the span l_r, and the numbers c and d depending only upon the lengths and number of the spans of the girder.* To find their numerical value for any given case is hence a simple arithmetical exercise.

Example 1.—A continuous girder has four unequal spans, $l_1=80$ ft., $l_2=100$ ft., $l_3=50$ ft., and $l_4=40$ ft. (Let the reader draw the figure). On the span l_2 is a single load $P=10$ tons, whose distance from the support 2 is $k\,l_2=40$ ft. To find the moments at the supports.

Since $k\,l_2=40$, and $l_2=100$, we have $k=0.4$. Inserting then in (I), the values of k, l_2 and P, we find

$$A=38400 \text{ tons ft.} \quad B=33600 \text{ tons ft.}$$

Inserting next the lengths of the spans in (II), we have

$$\begin{array}{ll} c_1=0 & d_1=0 \\ c_2=1 & d_2=1 \\ c_3=-3.6 & d_3=-3.6 \\ c_4=19.6 & d_4=10.3 \end{array}$$

* In the *London Philosophical Magazine* for September, 1875, where the above demonstration was first giver, the author has extended the method to girders witu fastened or walled-in ends.

Since the load is in the second span $r=2$, also $s=4$; hence $d_{s-r+2}=d_4=10.3$, $c_r = c_2=1$, etc., and $l_{s-1}=l_3=50$, etc. By inserting these values of c, d, l, A and B in (III), we obtain

$$M_n = \quad 82.01\ c_n, \qquad \text{when } n<3$$
$$M_n = -24.65\ d_{6-n}, \quad \text{when } n>2$$

For the abutment or left hand support, we have $n=1$, $c_1=0$, and hence $M_1=0$, for the second support, $n=2$, $c_2=1$, and $M_2=82.01$ tons ft. For the third support, $n=3$, $d_{6-n}=-3.6$, and $M_3=88.56$ tons ft. For the next, $n=4$, $d_{6-n}=1$, and $M_3=-24.65$ tons ft. Lastly, for the right hand abutment, $n=5$, and $M_5=0$. A positive moment, it will be remembered, causes tension in the upper chord of a truss, while a negative moment causes the reverse.

2. A girder of four equal spans has a load P at any point on the first span. Find the moment at each pier due to P.

Ans. $M_3=\frac{15}{56}Pl\,(k-k^3)$, etc.

SHEARS AND REACTIONS AT THE SUPPORTS.

It is thus easy from (I), (II) and (III) to find the moments at all supports due either to single concentrated loads or to a uniform load over an entire span, and these are the only two kinds of loading which we need to consider in designing a continuous bridge. We next need the shear S_n at the right of any support due to these same loads.

In computing strains in a continuous truss we take up each span separately. The index n refers always to the particular span under consideration, while the index r referring to the span in which the load is for the moment considered, may be less than, equal to or greater than n. For single loads the shear S_r is given directly by (3), for a uniform load $P(1-k)$ in that expression becomes $\frac{1}{2}wl_r$, while for an unloaded span those terms disappear. Thus we have

$$S_r = \frac{M_r - M_{r+1}}{l_r} + P(1-k), \text{ for Fig. 6.}$$

$$\text{(IV)} \quad S_r = \frac{M_r - M_{r+1}}{l_r} + \tfrac{1}{2} w\, l_r, \text{ for Fig. 3.}$$

$$S_n = \frac{M_n - M_{n+1}}{l_n}, \text{ when } n > r, \text{ or } n < r,$$

for the shear in the span l_n infinitely near to the support n.

The shear in the span l_n at a point infinitely near to the $n+1^{th}$ support is called S'_n (see Fig. 3). For its values we deduce

$$S'_r = \frac{M_{r+1} - M_r}{l_r} + P\,k, \text{ for Fig. 6.}$$

$$S'_r = \frac{M_{r+1} - M_r}{l_r} + \tfrac{1}{2}\,w l_r, \text{ for Fig. 3.}$$

$$S'_n = \frac{M_{n+1} - M_n}{l_n}, \text{ for } n > r, \text{ or } n < r.$$

The reaction R_n at any support n is evidently the sum of the shear S_n in the span l_n; and of the shear S'_{n-1} in the span l_{n-1} or for all cases

$$R_n = S_n + S'_{n-1}.$$

Example.—A girder of four equal spans has a single load P at the center of the third span from the left end. Find the reaction at the third support.

Ans. $R_3 = \frac{17}{28}$ P.

SHEAR AND MOMENT AT ANY SECTION.

These are given directly by the simple conditions of static equilibrium. For a single load we have from (2), (see Figs. 5 and 6),

(V) $S = S_r - P$, for a section between P and $r+1$
$S = S_n$, for any other section.

as expressing the internal shear for any section x (see Figs. 5 and 6).

Also, we have

(VI) $M = M_r - S_r x + P(x - k l_r)$, between P and $r+1$
$M = M_n - S_n x$, for any other section,

for the internal moment for any section x in a span either unloaded or containing the weight P. Similar expressions may be also written as in Chapter I, if the load be taken as uniformly distributed.

Example.—A girder of three equal spans has a load P at the center of the first spans. What is the shear and moment at the center of the middle span?

Ans. $S = \frac{1}{8} P$, $M = \frac{3}{80} Pl$.

MAXIMUM SHEARS AND MOMENTS.

The formulae (I) to (VI) inclusive are sufficient in connection with an arithmetical process of tabulation to determine the maximum strains in all the pieces of a properly designed continuous truss. Having for instance to calculate tbe span l_n, we may take at every panel apex throughout the bridge a single load P, and compute the shear and moment at *any* section due to every possible position of P. These arranged in a table, afford a clear view of the distribution of loading giving the maxima ; the greatest positive shear, for example, occurring when the live load covers those portions of the bridge which furnish plus values of S, while at the same time, it is absent from those portions giving minus values of S. Adding then all the plus values thus found, the maximum is determined by combination with that due to the always existing dead load.

It is therefore not absolutely necessary that the engineer should be acquainted with the theory of the distribution of loading giving rise to the maximum strains in the various pieces of the truss. As such a knowledge, however, is of great assistance in checking the accuracy of the calculation, we shall here state without demonstration the cases under which such maxima and minima arise.

First the shear; from this we obtain the strains in the webbing by the simple multiplication by a constant, a positive shear producing tension in a diagonal which slopes upward toward the left hand support. The maximum positive shear in the span l_n at the section whose distance from the support n is x, occurs under a distribution of loading such as is represented in Fig. 7, in which the shaded portions denote the live or rolling load. From this we see that the nearest span on the left and each alternate one

FIG. 7.

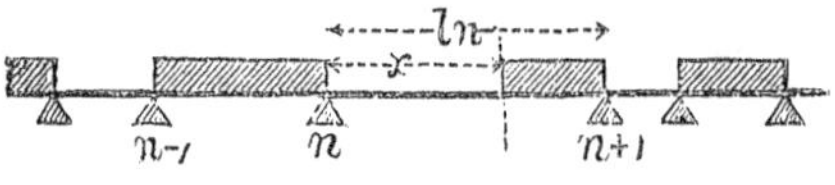

l_n x $n-1$ n $n+1$

are covered with the live load; that from the section x to the support $n+1$ the live load extends; and that the second span on the right and each alternate one are also covered with the live load; all other portions being subjected only to the dead or actual load of the truss. The minimum positive, or, what is the same thing, the maximum negative shear obtains under exactly reverse conditions, the loaded portions in Fig. 7 being unloaded, while the empty ones receive the live load. Let the reader draw a figure for this case, and imagine the section x to move from n to $n+1$.

Next the moment; from this we obtain the chord strains by dividing by the constant depth of the truss, a positive moment producing tension in the upper chord. Here the maximum positive moment in the span l_n, occurs near the support n under a distribution of loading like that represented in the first illustration of Fig. 8, near the middle of the span as in the second and near the support $n+1$ as in the last. Fig. 8 represents one and the same beam with the cases of loading causing maximum positive moments at three different sections in the span l_n, the first a section between n and a point i, the second between i and i' and the third between i' and $n+1$. These points i and i' are called *fixed inflection points*, and they enjoy the property that all loads on the spans to the

FIG. 8.

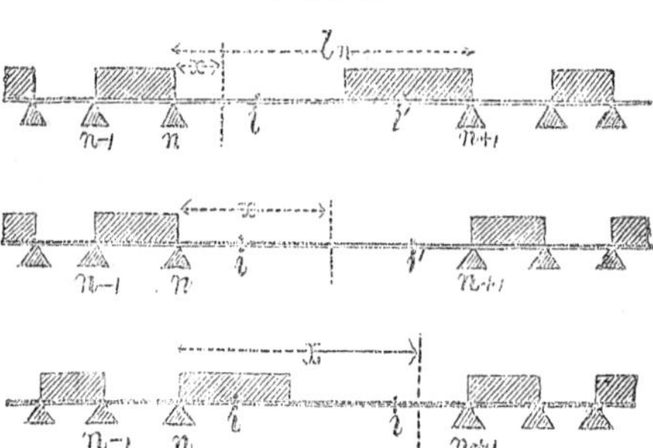

right of l_n, produce no moment at i, while all loads on the spans to the left of l_n produce no moment at i'. The position of these points depend only upon the lengths and number of the spans of the girder, or upon the numbers c and d given by (II). If the distance from n to i be denoted by i, and that from n to i' by i', the following simple formulæ

$$i = \frac{c_n l_n}{c_n - c_{n+1}}$$

$$i' = \frac{d_{s-n+2} l_n}{d_{s-n+2} - d_{s-n+1}}$$

will give the position of the fixed inflection points in any span l_n.

In order to render these distributions of load clear, let us imagine the section x to move from the support n to $n+1$. When the section is at n the live load covers the whole span l_n to render the moment at x a maximum, as x passes toward i the load recedes rapidly toward $n+1$, until when x reaches i the span l_n becomes empty, and the loads on the following spans shift as shown in Fig. 8. As x passes from i to i' the span l_n remains empty as in the second sketch aud when it reaches i' the loads on the preceding spans shift. As soon as x passes i' the load begins to come on at n, which rapidly increases as x moves, until it covers the whole span when x coincides with $n+1$.

The arrangements of loading for causing the maximum negative moment in any section depend likewise upon the position of that section with reference to the fixed inflection points, and are in all cases exactly the reverse of those for the positive moment.

It will be seen, then, that the maximum moments between the supports and the fixed inflection points cannot be determined by cases of loading, for such cases are different for every section. In a girder of two equal spans for example, one of these points in each span coincides with the abutments, the others are at one-fifth the length of the span from the pier. Between those points the maximum strains are not to be found by parabolic curves of moments drawn from a few assumed arrangements of loading. Here have some late writers fallen into grave error.

The above completes, what seems to

us a simple presentation of the theory of the continuous girder of constant cross-section. We have disconnected it entirely from the properties of the simple girder, have avoided the use of artificial angles and couples, parabolic moment and shear curves, static and elastic reactions and other paraphernalia which are too often introduced to complicate the subject. The formulae (I) to (VI) which may be written on a page of the engineer's note book include indeed the whole theory, and are sufficient for the determination of the maximum strains in a continuous truss of any number or lengths of spans.

CHAPTER III.

We will now apply the above theory to the practical calculation of the strains in a continuous truss, and to show the perfect generality of our method we will take one of *five unequal spans.*

Fig. 9 shows the relative lengths of the several spans, each support and span receiving an index according to the notation previously adopted. The first span l_1 has a length of 70 feet, the

FIG. 9.

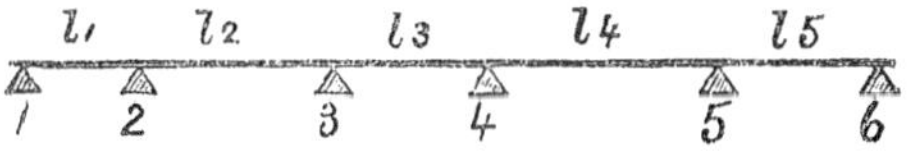

l_1 l_2 l_3 l_4 l_5

1 2 3 4 5 6

second l_2 of 100 feet, the third l_3 of 80 feet, the fourth l_4 of 120 feet and the last l_5 of 90 feet. This girder is to be subject to a live load of 0.8 tons per linear foot per truss ; the dead load we estimate at 0.6 tons per linear foot per truss. It is divided into panels of ten feet each and its height is also ten feet, the webbing being a simple series of isosceles triangles as shown in Fig. 10, which represents the

FIG. 10.

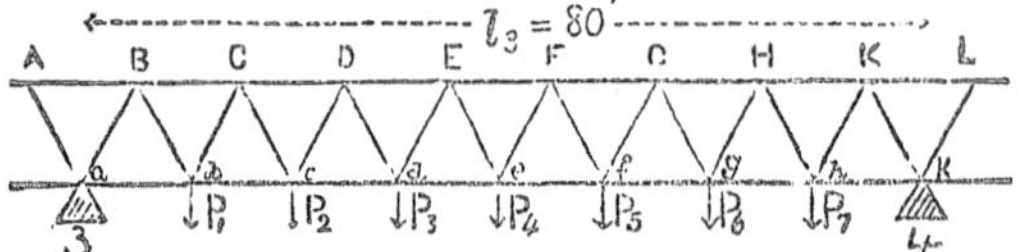

span l_3 enlarged. The live load is applied at the panel points on the lower chord. It is required to calculate the maximum strains in all the pieces of the span l_3 due to the above live and dead loads.

We take up first the live load of 0.8 tons per linear foot, or eight tons per panel. Since every load in the span l_1 affects every section in l_3 in a similar manner, we may, instead of considering the separate panel loads on l_1, take them as uniformly distributed in the preliminary determination of the shear and moment at 3. (The load of eight tons at the points 1, 2, 3, etc., give reactions only at those points, and cannot affect the span l_3). On the span l_1 there are seven panels and six apices, hence the live load in that span is $W_1=6\times8=48$ tons. In the same way we have on the spans l_2, l_4 and l_5, to consider the live loads applied at the panel apices as uniformly distributed over the spans; but in the span l_3 we must consider each panel load separately, since different arrangements of those loads give maxima for different sections. Thus we have

On l_1, the load $W_1=48$ tons,
On l_2, the load $W_2=72$ tons,
On l_3, the loads P_1, P_2, P_3, etc.,

(see Fig. 10) each equal to 8 tons,

On l_4, the load $W_4=38$ tons,
On l_5, the load $W_5=64$ tons.

We now turn to formulae (I) of the

preceding chapter, and determine the quantities A and B due to each of these loads. For that on l_1 we have, for example, $W_1=48$, $l^2_1=4900$, hence $A=B=58800$. For P_1 we have $P=8$, $l^2_3=6400$, $k=\frac{1}{8}$, hence $A=10500$, and $B=6300$; in like manner, for P_2, P_3, etc., we place $k=\frac{2}{8}$, $k=\frac{3}{8}$, etc., and find for each a value of A and B. Thus we have for the several loads :

For W_1,	$A=B=$ 58800	
For W_2,	$A=B=$180000	
For P_1,	$A=10500$	$B=$ 6300
For P_2,	$A=16800$	$B=12000$
For P_3,	$A=19500$	$B=16500$
For P_4,	$A=19200$	$B=19200$
For P_5,	$A=16500$	$B=19500$
For P_6,	$A=12000$	$B=16800$
For P_7,	$A=$ 6300	$B=10500$
For W_4,	$A=B=$316800	
For W_5,	$A=B=$129600	

We next turn to formulae (II), and substitute the lengths $l_1=70$, $l_2=100$, etc., and thus obtain

$c_1=0$	$d_1=0$
$c_2=1$	$d_2=1$
$c_3=-3.4$	$d_3=-3.5$
$c_4=14.05$	$d_4=16.$
$c_5=-44.567$	$d_5=-54.8$

as the values of the multipliers c and d for the case under consideration.

We are now able to find from (III) the moments M_3 and M_4 at the supports 3 and 4 for each of the above loads. Since there are five spans, $s=5$, $d_s=-54.8$, $d_{s-1}=16$, etc.; substituting these in (III), we have

$$M_n = -c_n \frac{A\, d_{7-r} + B_{6-r}}{17032}, \text{when } n<r+1$$

$$M_n = -d_{7-n} \frac{A\, c_r + B\, c_{r+1}}{17032}, \text{ when } n>r$$

n being *any* index (in our case either 3 or 4), and r the index of that span, which, for the moment, we regard as loaded. Taking the load on the first span, we have $r=1$, and since $n>r$, we use only the second of the above formulae, which becomes

$$M_n = -d_{7-n}\frac{A c_1 + B c_2}{17032} = -3.452\ d_{7-n}$$

Substituting in this $n=3$ and $n=4$, and we have

$$M_3 = -3.452\ (16) = -55.23 \text{ tons ft.}$$
$$M_4 = -3.452\ (-3.5) = 12.08 \text{ tons ft.}$$

Taking the load in the second span, we have $r=2$, and

$$M_n = -d_{7-n}\frac{A c_2 + B c_3}{17032} = -25.364\ d_{7-n}$$

in which, by making $n=3$ and $n=4$, we get the values of M_3 and M_4.

For the single loads on the span l_3, we must use the first of formulae (III) to obtain M_3, and the second to obtain M_4. Making then $r=3$, we have,

$$M_3 = -c_3\frac{A d_4 + B d_3}{17032} = \frac{3.4}{17032}$$
$$(16\,A - 3.5\,B)$$
$$M_4 = -d_3\frac{A c_3 + B c_4}{17032} = \frac{3.5}{17032}$$
$$(-3.4\,A + 14.05\,B)$$

and inserting in these the values of A

and B, as given above, we find the moments at 3 and 4 from each of the loads from P_1 to P_7. For the load on the fourth span, we make $r=4$; for that on the fifth span, $r=5$; and the first of formulae (III) give the moments. Thus, by very simple arithmetical work, we obtain the moments M_3 and M_4 due to each of the single loads in l_3, and each of the uniform loads in the exterior spans, and arrange them in the second and third columns of the following table :

Load.	M_3	M_4	S_3
	tons ft.	tons ft.	tons.
W_1	− 55.23	+ 12 08	− 0.841
W_2	+405.82	− 88.77	+ 6.182
P_1	+ 31.19	+ 10.83	+ 7.254
P_2	+ 45.36	+ 22.85	+ 6.281
P_3	+ 50.85	+ 33.93	+ 5.211
P_4	+ 48.00	+ 41.92	+ 4 078
P_5	+ 39.15	+ 44.65	+ 2.931
P_6	+ 26.64	+ 39.92	+ 1.834
P_7	+ 12.81	+ 25.85	+ 0.837
W_4	−158.10	+653.34	−10.143
W_5	+ 25.87	−106.91	+ 1.660

The last column of the table, which gives the shear S_3 in the span l_3 at a point infinitely near to the support 3, is found from the quantities M_3 and M_4 by means of formulae (IV). The load W_2, for example, gives a positive moment of 405.82 tons ft. at 3, and a negative one of 88.77 tons ft. at 4. From the last formula of (IV), we have then

$$S_3 = \frac{405.83 + 88.77}{80} = 6.182 \text{ tons.}$$

Also for the load P_3 on the span l_3, we have $P=8$, $k=\frac{3}{8}$, and from the first formulae of (IV)

$$S_3 = \frac{50.85 - 33.93}{80} + 8\left(1 - \tfrac{3}{8}\right) = 5.211 \text{ tons.}$$

and in the same way the other shears in the last column are computed. All of these refer, of course, only to the live load of eight tons per panel.

We are now ready to proceed with the computation of the maximum strains in the span l_3. And first we take up the webbing.

The maximum strain in any diagonal in Fig. 10, is equal to the maximum shear for that section multiplied by the secant of the angle between the diagonal and a vertical. We proceed first to find the maximum shears.

The shear at any section due to the dead load is constant, increases or decreases as the live load comes upon the bridge, and becomes a maximum or minimum under certain particular distributions of loading. To determine these it is only necessary to tabulate the shear due to each separate load. This is easily done from the values of S_3 and the formulae (V). In the following table the vertical column headed *aBb* includes the shears which may act upon the diagonals *a* B and B *b*, *b* C *c* those for *b* C and C *c*, etc. The horizontal column numbered 1 gives then the shears at every section due to the live loads; the load W_1 for example producing a negative shear of 0.84 tons in every panel or $S=S_3$, the load P_3 giving in the three panels on its left $S = S_3 = +5.21$ tons and in the five on its right $S-S_3-P= +5.21-8= -2.79$ tons. A mere inspection of this table

Shears. (See Fig. 10.)

		a B *b*.	*b* C *c*.	*c* D *d*.	*d* E *e*.	*e* F *f*.	*f* G *g*.	*g* H *h*.	*h* K *k*.
1	W_1	−0.84	−0.84	−0.84	−0.84	−0.84	−0.84	−0.84	−0.84
	W_2	+6.18	+6.18	+6.18	+6.18	+6.18	+6.18	+6.18	+6.18
	P_1	+7.25	−0.75	−0.75	−0.75	−0.75	−0.75	−0.75	−0.75
	P_2	+6.28	+6.28	−1.72	−1.72	−1.72	−1.72	−1.72	−1.72
	P_3	+5.21	+5.21	+5.21	−2.79	−2.79	−2.79	−2.79	−2.79
	P_4	+4.08	+4.08	+4.08	+4.08	−3.92	−3.92	−3.92	−3.92
	P_5	+2.93	+2.93	+2.93	+2.93	+2.93	−5.07	−5.07	−5.07
	P_6	+1.83	+1.83	+1.83	+1.83	+1.83	+1.83	−6.17	−6.17
	P_7	+0.84	+0.84	+0.84	+0.84	+0.84	+0.84	+0.84	−7.16
	W_4	−10.14	−10.14	−10.14	−10.14	−10.14	−10.14	−10.14	−10.14
	W_5	+1.66	+1.66	+1.66	+1.66	+1.66	+1.66	+1.66	+1.66
2	Live load +	+36.26	+29.01	+22.73	+17.52	+13.44	+10.51	+8.68	+7.84
	Live load −	−10.98	−11.73	−13.45	−16.24	−20.16	−25.23	−31.40	−38.56
3	Sums	+25.28	+17.28	+9.28	+1.28	−6.72	−14.72	−22.72	−30.72
4	Dead load	+18.96	+12.96	+6.96	+0.96	−5.04	−11.04	−17.04	−23.04
5	Max. +	+55.22	+41.97	+29.69	+18.48	+8.40			
	Max. −			−6.49	−15.28	−25.20	−36.27	−48.44	−61.60

shows the distribution of live load causing the maximum or minimum shear in any section. Thus for the panel *d* E *e* the maximum occurs when those loads giving positive shears are present, namely, W_2, P_4, P_5, P_6, P_7. and W_8 and when all the others are absent, and the minimum occurs when only those giving negative shears are on the bridge. If then we add all the positive quantities in 1 and likewise all the negative ones and place the results in the horizontal column 2, we have for the panel *d* E *e*, +17.52 tons and −16.24 tons as the greatest and least shears which can occur in that panel due to the live load, and these need only to be combined with the shear due to the dead load to obtain the absolute maximum and minimum.

If the dead load be regarded like the live load as concentrated at the panel points on the lower chord, its effect will be a fractional part of that of the live considered as uniformly distributed. Adding algebraically the quantities in 2 we have in 3 the shears produced by a uniformly distributed live load of eight tons per panel, since this is the same as taking the algebraic sum of all the quantities in 1. The live load if extending over the whole bridge will then produce in *d E e* a shear of +1.28 tons, and since the actual dead load is three-fourths of the live, the dead load must produce in that panel a shear equal to $\frac{3}{4}\times 1.28 = 0.96$ tons. Taking then three-fourths of the quantities in the horizontal column 3 we have in 4 the shears due to the dead load of six tons per panel.

The shears in 4 always *must* exist, while those in 2 *may* exist under certain positions of the live load. The absolute maxima are therefore found by adding algebraically the quantities in those two horizontal rows. Thus for *d E e*, +0.96 always obtains, and if +17.52 also occurs their sum +18.48 is the positive maximum shear; and if −16.24 occurs, $+0.96 - 16.24 = -15.28$ is the minimum or negative maximum. Placing these results in column 5 we have the required maximum shears in every section of the span under consideration. If the dead load shear in 4 is greater than the live load shear of opposite sign in 2 only one kind of shear can prevail; thus in *b* C *c* the greatest possible value is +12.96 +29.01 = +41.97 tons, the least possible is +12.96 −11.73 = +1.23 tons and the diagonals *b* C and C *c* will be subject to only one kind of strain. In the case before us three panels *c* D *d*, *d* E *e* and *e* F *f* have both a positive and negative maximum and the diagonals in those panels may be subject to either tension or compression.

The maximum shears in 5 multiplied by *sec.* θ, or the secant of the inclination of diagonal to vertical give the maximum strains, tension of the diagonal slopes upward toward the left hand support, compression if it slopes downward. The depth of the truss being ten feet and the half panel length 5 feet, *sec.* θ is 1.118. We have then the following table of maximum strains in the

DIAGONALS. (See Fig. 10.)

B*a*	—61.7 tons
B*b*	+61.7
C*b*	—46.9
C*c*	+46.9
D*c*	—33.2 or +7.2 tons
D*d*	+33.3 or —7.2
E*d*	—20.6 or +17.1
E*e*	+20 6 or —17.1
F*e*	+28.2 or —9.4
F*f*	—28.2 or +9.4
G*f*	+40.5
G*g*	—40.5
H*g*	+54.0
H*h*	—54.0
K*h*	+68 3
K*k*	—68.3

in which + denotes tension and — compression.

In the same manner we may tabulate the moments at every section due to each load and deduce the maximum chord strains. For the upper chord panels A B, B C, C D, etc., the centers of moments are at the opposite vertices *a*, *b*, *c*, etc., and hence in formulae (VI) we must take 0, 10, 20, etc., as the successive values of x. To find the moment for CD due to the load W_1 on the span l_1, we have only to insert $x=20$, and the values of M_3 and S_3 as found above for that load (see Fig. 10), giving

$$M = -55.23 - (-0.841 \times 20)$$
$$= -38.41 \text{ tons ft.}$$

the negative sign denoting that W_1 causes compression in the upper chord. Also to find the moment in the same panel due to the load P_1, we have as before $x=20$ and

$$M = 31.19 - 7.254 \times 20 + 8 \times 10$$
$$= -33.89 \text{ tons ft.}$$

In this way we readily compute the moments for every panel due to every live load and arrange them as in the following table of

MOMENTS FOR UPPER CHORD. (See Fig. 10.)

		A B.	B C.	C D.	D E.	E F.	F G.	G H.	H K.	K L.
1	W_1	—55.2	—46.8	—38.4	—30.0	—21.6	—13.2	—4.8	+3.6	+12.1
	W_2	+405.8	+344.0	+282.2	+220.4	+158.5	+96.7	+34.9	—26.9	—88.8
	P_1	+31.2	—41.4	—33.9	—26.4	—19.0	—11.5	—4.1	+3.4	+10.8
	P_2	+45.4	—17.5	—80.3	—63.1	—45.9	—28.7	—11.3	" 5.7	" 22.9
	P_3	+50.9	—1.3	—53.4	—105.4	—77.6	—49.7	—21.8	" 6.1	" 33.9
	P_4	+48.0	+7.2	—33.6	—74.3	—115.1	—75.8	—36.6	" 2.6	" 41.9
	P_5	+39.2	+9.8	—20.5	—49.8	—79.1	—108.4	—56.7	—6.6	" 44.7
	P_6	+26.6	+8.3	—10.6	—28.4	—46.7	—65.1	—83.4	—21.7	" 39.9
	P_7	+12.8	+4.4	—3.9	—12.3	—20.7	—29.0	—37.4	—45.8	" 25.9
	W_4	—158.1	—56.7	+44.8	+146.9	+247.6	+349.1	+450.5	+551.9	+653.3
	W_5	+25.9	+9.3	—7.3	—23.9	—40.5	—57.1	—73.7	—90.3	—106.9
2	Live load +	+685.8	+383.0	+327.0	+367.3	+406.1	+445.8	+485.4	+573.3	+885.4
	Live load —	—213.3	—163.7	—281.3	—413.6	—466.2	—438.5	—330.0	—190.7	—195.7
3	Sums	+472.5	+219.3	+45.7	—46.3	—60.1	+7.3	+155.4	+382.6	+689.7
4	Dead load	+354.3	+164.4	+34.2	—35.4	—45.0	+5.4	+116.4	+286.8	+517.2
5	Max. +	+1040.1	+547.4	+361.2	+331.9	+361.1	+450.2	+601.8	+860.1	+1402.6
	Max. —			—247.1	—449.0	—411.2	—433.1	—213.6		

The horizontal column 1 of this table shows at a glance the distribution of live load giving the maximum strain in any bay: in the bay B C for instance the loads W_2 W_5 and P_4 to P_7 inclusive produce positive moments, and hence the greatest tensile strain obtains when those loads are on the bridge and all others are absent, while the least tensile strain in B C occurs when W_1, W_4, P_1, P_2 and P_3 are present and the others absent. Adding separately therefore these positive and negative moments, we have in 2, the greatest and least moments for every bay due to the live load. Adding those algebraically and we have in 3 the moments when the live load covers the entire bridge; taking three-fourths of the quantities in 3 we have in 4 the moments due to the actual dead load. Lastly combining the moments in 4 which always *must* exist with those in 2 which *may* exist, we get in 5 the absolute maxima and minima. For example, in C D we have due to the live load the moments +327.0 and −281.3, the sum of these or +45.7 is the moment when the live load extends over the whole girder, three-fourths of this or +34.2 is the value due to the actual dead load, and finally

$$+34.2+327.0=+361.2 \text{ tons ft.}$$
$$+34.2-281.3=-247.1 \text{ tons ft.}$$

are the maximum positive and negative moments, C D may then be subject to two kinds of strain.

Dividing these results by the depth of the truss and remembering that a positive moment causes tension in the upper chord, we have the maximum strains for the

UPPER CHORD. (See Fig. 10.)

A B	+104.0 tons
B C	+ 54 7
C D	+ 36.1 or —24.7 tons
D E	+ 33.2 or —44.9
E F	+ 36.1 or —41 1
F G	+ 45.0 or —43.3
G H	+ 60.2 or —21.4
H K	+ 86 0
K L	+140.3

From this we see that the whole upper chord may under certain positions of the rolling load be subject to tension. This is due to the short length of the span compared with the adjacent ones.

The calculation of the maximum strains in the lower chord is entirely similar, the centers of moments being at the opposite vertices B, C, D, etc., and the corresponding values of x being 5, 15, 25, etc. We leave therefore as an exercise for the student the formation of the tabulation, merely giving the results to which it will lead, viz.

Lower Chord. (See Fig. 10.)

a b	— 76.9 tons
b c	— 41.4
c d	— 34.7 or + 34.6 tons
d e	— 34.7 or + 47 8
e f	— 40.7 or + 47.1
f g	— 51.9 or + 34.0
g h	— 70 7
h k	—127.9

The strain sheet for the span l_3 (Fig.9) is now finished, and in a similar way each of the other spans may be computed. The method we have presented is entirely general and applicable to any number of continuous spans, whether equal or unequal. In each case we take the load in the exterior spans as uniform and that in the span under consideration as applied at the panel apices, and find for each the quantities A and B from (I) and from the lengths of the spans we find by (II) the multipliers c and d. These enable us to deduce from (III) the moments at the supports due to each load, from which (IV) give us the shear. It is only in this preliminary computation of moments and shears that the calculation of continuous girders differs from that of ordinary simple trusses. In the latter the moments at the ends are known to be zero, and the shears coincide with the reactions which are found from the law of the lever. The simple truss is thus but a particular case of the continuous one as may be readily seen by placing $s=1$ in our formulae (I) to (IV). In an *end* span of a continuous truss the moment at the abutment is also zero and the shear is the same as the reaction.*

GIRDERS WITH SPANS ALL EQUAL.

This is one of the most common cases. Making in (II) all the l's equal, we have

$$c_1 = d_1 = 0$$
$$c_2 = d_2 = 1$$
$$c_3 = d_3 = -4$$
$$c_4 = d_4 = 15$$
$$c_5 = d_5 = -56$$
$$c_6 = d_6 = 209, \text{ etc.,}$$

which are the well-known Clapeyronian numbers first deduced by the discoverer of the theorem of three moments.† Each of these numbers is equal to four times the preceding one less the one next preceding, and their signs are alternately positive and negative. They are here seen to be a particular case of our general formulae (II).

* For an example of the computation of a continuous truss of two spans, see Van Nostrand's *Engineering Magazine* for July, 1875.

† See *Comptes Rendus*, 1857, p. 1076.

If the two end spans of the bridge are each equal to βl and the others each equal to l, the multipliers c and d become also equal. Their values are

$$c_1 = d_1 = 0$$
$$c_2 = d_2 = 1$$
$$c_3 = d_3 = -2 - 2\beta$$
$$c_4 = d_4 = 7 + 8\beta$$
$$c_5 = d_5 = -26 - 30\beta$$
$$c_6 = d_6 = 97 + 112\beta, \text{ etc.,}$$

each being equal to four times the preceding one less the one preceding that. Having established the value of β (usually taken at about 0.8) these reduce at once to numerical form. If β be unity the spans become all equal and the multipliers reduce to the Clapeyronian numbers.

Example.—A girder of four spans has a single load P in the second span at a distance kl from the second support; the two end spans being equal to $0.8\,l$ and the central ones to l. Find the moment at the second support.

Ans. $M_2 = Pl(0.52k - 0.901\,k^2 + 0.381k^3)$

Having computed the maximum strains in a continuous truss, we choose for the various pieces cross sections of an area and form sufficient to resist those strains, thus making the girder one of uniform strength. The theory by which we have computed the strains supposes however that the cross section of the girder is constant. The question now arises what error is introduced by this hypothesis.

As we have been unable to present in the short limits of this paper the theory of the continuous girder with variable cross section, we cannot place before the reader a mathematical comparison of the two cases, and are hence obliged to rely on the computations of others and on general considerations.

Computations of strains in continuous girders have been made by Bresse, Mohr, Winkler, Weyrauch and others, considering the cross section both constant and variable. The general conclusion to be derived from their investigations is, that the maximum moments over the supports are greater, when the variable cross section is taken into account, but rarely more than six per cent. that the maximum moments near the centers of the spans are generally slightly less, and that the shearing forces do not sensibly differ.* For example, in a truss of two equal spans, the maximum moment at the pier is 0.125 wl^2 for constant cross section and 0.133 wl^2 for variable ; the maximum negative moment is 0.070 wl^2 for constant and 0.067 $w\,l^2$ for variable, and the reactions of the pier are 1.250 $w\,l$ and 1.266 $w\,l$ respectively.

If then we compute continuous trusses as if they were of constant cross section, we are liable to slight errors in the chord strains. These strains are however computed on the assumption of a distribution of live load which can never occur in practice, and in proportioning the sec-

---

* Mohr. in *Zeitschrift des Arch. u. Ing. Ver. zu Hannover*, 1860, 1862. Winkler; *Die Lehre von der Elasticität*, p. 150. Weyrauch; *Theorie der continuirlichen Träger*, p. 22., p. 143.

tions to those strains a factor of safety involving five or six-fold security is introduced. Considering then that it must be almost impossible for the live load to be arranged on the bridge as Fig. 8 represents, we may be well assured that our computations on the hypothesis of constant moment of inertia give greater strains than can ever obtain.

After having computed on both hypotheses a girder with four spans, two of sixty-five meters in length and two of fifty-two meters, Weyrauch says: "We are now able to answer the question, whether it is allowable to calculate continuous girders with variable cross section by the formulae for constant cross section, in the affirmative. The maximum moments arising from the two calculations differ but slightly. In our example, where the cross sections vary between 1 and 2⅓, the greatest difference was 6 per cent. the next following only 2.7 per cent. The shears change scarcely at all. Only for bridges with extremely long spans, is it desirable to make a second computation on account of variable cross section."

LEVELS OF SUPPORTS.

In Chapter I we alluded to the fact that small changes in the relative levels of the piers produce great variations in the strains of the pieces of the truss. Continuous girders should not for this reason be used when the piers are liable to settle. Whether the supporting points are exactly upon level when the bridge is built is a matter of no importance, although, of course, no great differences can be allowed.

If in finding the equation of the elastic line we had considered the supports on different levels, a term containing those differences of level and the term E I would have entered the theorem of three moments. Now if a straight beam be laid across two points, a downward force is necessary in order that it should touch a third point at a lower level. If this downward force be furnished by the weight of the beam, a certain part of this weight will be effective in producing the deflection or satisfying the term E I, while the remaining part will act exactly as if the three supports were on the same level. But if the beam, instead of being originally straight, were of such a shape that it exactly fitted the three points it is in the same condition as the horizontal one after it has undergone the deflection. Hence its action is independent of the variations in level, for no exterior force is required to compel it to correspond with the points of support.

We therefore conclude, that if a bridge be built, by suitably adjusting its false works, corresponding to the profile of the piers, all the strains obtain exactly as if those piers were on one and the same level.

BEST LENGTHS OF SPANS.

Except the well-known rule that the lengths of the spans should be so adjusted that the cost of the piers and superstructure may be equal, there is little to be said upon this subject. Although most writers give a mathematical discussion of the most economical relations of spans, the fact that no two of them agree except in the simplest case, only indicates that the theory contains no principle which will lead to general conclusions.

Viewing the matter from a practical point of view, but not neglecting the investigations of mathematicians, we may give the following rules. For two spans the lengths should be equal. For others the span should be symmetrical, the interior ones being equal and the end ones shorter by about one-fifth or one-sixth, or making the central ones each equal to l, the end ones should be about $\frac{4}{5} l$ or $\frac{5}{6} l$. Such an arrangement equalizes the moments due to the dead load and being pleasing to the eye, it is advisable to regard it in designing continuous bridges.

PRACTICAL CONSTRUCTION.

In the construction of continuous bridges, the following points should be carefully observed :

1. The iron should be of a uniform qnality, and have undergone as near as possible the same process of manufacture.

2. The truss should be built with parallel chords, each capable of resisting either tension or compression; the webbing should be simple. With double and triple systems of webbing the strains cannot be accurately determined.

3. Joints in the chords should never be made over the piers.

4. The false works should be so adjusted that the bridge may be built with its points of support on the same relative level with the actual bed plates.

5. The bearing surface of the bridge upon the piers should be as small as possible consistent with considerations of strength and safety, and arrangements for longitudinal variation, due to changes in temperature, should be provided.

6. Continuous girders cannot be used if the piers are liable to variations in level after erection.

ADVANTAGES OF THE CONTINUOUS SYSTEM.

In favor of the continuous girder *versus* the simple one, we may mention :

1. Greater stiffness, since the deflection under a rolling load is much less than that of independent simple spans.

2. Ease of erection in cases where false works are difficult and expensive ; the girder may then be built on shore and pushed out over the piers.

3. Saving in material for the piers, since a less bearing surface is required than for two ends of single span bridges.

4. *Saving in iron, amounting to from twenty to forty per cent. over the ordinary construction of single spans.*

5. Simplicity of construction, when an angle of skew exists in the piers: in such cases the cross girders may be placed at right angles in the continuous structure, and the difficulties of oblique connections entirely avoided.

In a simple girder, whose length is l, and live load per unit of length w, the maximum deflection due to the live load is $\frac{5\,w\,l^4}{384\,E\,I}$. In two continuous spans each equal to l, the maximum deflection, which occurs when one span is covered with the live load, is $\frac{3.7\,w\,l^4}{384\,E\,I}$, or only three-fourths as much. With many continuous spans, the deflection is much less, its greatest value being in the end spans. In the case of a girder with horizontally fastened ends, the deflection is only one-fifth of that of ordinary simple spans.

The saving in iron is large, and alone sufficient to recommend the continuous system, particularly for long spans. This saving occurs wholly in the chords where material can best be spared. In the webbing the quantity of material is slightly increased. The exact percentage of saving depends upon the number and lengths of spans, the proportion of live to dead load, the arrangement of webbing, and will be the same in no two cases. In the example of Fig. 9, the center span which we computed does not afford scarcely any saving in material owing to the influence of the larger adjacent spans. For girders of two hundred feet in length with spans nearly equal, calculation indicates a saving of about thirty per cent. For the extreme case—a single span with horizontal fixed ends—the saving is fifty per cent.

DISADVANTAGES OF THE CONTINUOUS SYSTEM.

In our last chapter we referred to an article by Charles Bender, C. E., which contains many ingenious arguments against the use of continuous bridges. There are, in fact, fifteen "conclusions" to which he is led and which may be seen by the reader on p. 109 of the current volume of *The Journal of the American Society of Civil Engineers.* Of these we will give a short abstract and append a running commentary.

1. The theoretical calculation of curves of moments, without consideration of proportions and details, is exceedingly

fallacious.—Other things being equal, the calculation of strains does furnish an estimate of amounts of material. The method of curves moreover is neither so accurate nor so quick as our process of tabulation.

2. This fallacy will be greater, if the theory stands upon false premises.—Undoubtedly.

3. The theory of continuity is fallacious and unreliable, because it supposes the coefficient of elasticity constant, whereas it has been shown that it may vary for wrought iron from 17,000,000 to 40,000,000 lbs. per square inch.—In our last chapter we have shown that the proper interpretation of the variability of E is, that different qualities of iron have different degrees of elasticity. The values of E, from which it is concluded that the laws of flexure are fallacious, were in fact found by the theory itself from the measured deflections of beams, and it is hence more than fallacious to regard them as condemning that theory.

4. With several diagonal systems the

strains in a continuous girder cannot be calculated, but only guessed.—A good objection and it applies also to simple girders and draw spans.

5. The theory needs a correction if the chords are made variable in cross section. —The amount of this correction we have indicated above.

6. The calculation of continuous trusses is excedingly tedious, and if generally introduced would greatly impede the business of bridge building in this country.—A conclusion to which those who know how to calculate decidedly object. For a construction costing half a million dollars, it matters little whether one day or one week be spent in computation, and if the one week saves ten per cent or more on the cost, it is certainly well employed.*

* In the discussion of this subject at the late Engineers' Convention, one of the speakers mentioned an instance where the strain sheet for a continuous revolving draw bridge was furnished for $40. As that strain sheet was made by the author of this paper it is not improper that he should remark that its price would have been considerably less had the computations been made by tabulation of apex loads, instead of by the consideration of cases of loading.

7. Pieces which resist two kinds of strain must be proportioned to resist the maximum tension plus the maximum compression.—Further experiments are much needed in this connection; if Woehler's conclusions are confirmed, pieces must be so proportioned and hence the percentage of saving lowered.

8. Continuous girders require very accurate workmanship.—Are we to infer that ordinary trusses do not?

9. The foundations and masonry of the piers must be of excellent quality.—Does not the same hold for the simple girder? See No. 11.

10. If the girders are built on shore and pushed out over the piers, additional computations must be made and extra pieces introduced.—The additional computations are of the simplest character and in many cases no extra pieces would be needed.

11. If improperly placed on their bed plates, greater strains arise than are contemplated, an inch difference in level producing great variations in strains.—

We have shown above that unless piers are liable to settle after the erection of the bridge, such differences in level produce no effect.

12. If one chord be protected from the heat of the sun the strains are much disturbed.—The camber of the bridge is altered, as also occurs in single spans. It should be remembered however that although certain strains cause a curvature, a certain curvature does not necessarily cause corresponding strains.

13. Continuous bridges have proved to be more economical in Europe than simple spans, because the latter have been improperly proportioned.—Trial is necessary to prove that they would or would not be more economical in this country.

14. By designing two bridges of 200 feet each, a two span continuous truss twenty-five feet high, and a simple truss 27 feet high, with different details, Mr. Bender finds that the latter is more economical.—Perhaps for other proportions this might be reversed. In a com-

parison of amount of material, ought not other things to be taken equal?

15. Continuous bridges deflect as much as single spans.—Not if they have the same height and span, and are subject to the same loads; a fact which every schoolboy knows.

In regard to objections 2, 6, 7, 8, 9, 10, 12 and 13 one remark further is necessary. They are objections which may be made to every new proposed construction in engineering art. In the days of wooden bridges, they were advanced against the use of iron; they have been made against the suspension system and against the braced arch. But their value can be estimated in only one way, *by trial.* On the other hand theory can estimate one at least of the advantages claimed for the continuous systems, and that estimate is a saving of twenty to forty per cent. in material; how much of this must be deducted for extra care in workmanship, labor of erection, effects of temperature, or variations in the elasticity of the material can only be determined by the actual erection of continuous bridges, by experiments extending through a long series of years.

Having thus stated briefly, but fairly, the arguments for and against the use of continuous bridges, we leave it to practical builders to decide whether or not the system is worth a trial. Other nations have long been using it; profiting by their experience and by our own improved methods of manufacture and modes of erection, it may, perhaps, turn out that we shall find it better and more economical than the present system of single independent spans.

HISTORY AND LITERATURE.

The literature on the theory of continuous girders is very extensive in the German and French languages, and very limited in English. We can only give here a few hints concerning its development and history.

About the year 1825, Navier founded the present theory of flexure by introducing the hypothesis that the extensions and compressions of the fibers on each side of the neutral axis were proportional to their distances from that axis. From this he deduced the equation of the elastic line, and applied it to the discussion of continuous girders. His method consisted in determining first the reactions of the supports, and from these the internal forces or strains, which, although the most logical, was exceedingly tedious in practice. In the following works the reader may find detailed information concerning his method:

Kayser: *Handbuch der Statik*, Carlsruhe, 1836, chap. X.

Molinos et Pronnier: *Traité de la construction des ponts metaliques*, Paris, 1857.

From the time of Navier to the publication in 1857 by Clapeyron of the method of using the moments over the piers as auxiliaries in the computation instead of the reactions, many continuous girders were built in France, Germany and England. Among these may be mentioned the Britannia tubular bridge, of four spans, two of 231 feet, and two of 460 feet; the Boyne Viaduct with three spans of 141, 267 and 141 feet; and the bridge over the Weichsel at Dirschau with *six* continuous spans, each 397 feet in length. By Clapeyron's happy discovery of the theorem of three moments, a great impetus seemed to be given toward the erection of such bridges, for in the twenty years following 1857, we find them extensively built in Germany and France, although in England the unfortunate example of the Britannia tube caused a tendency to other forms of construction. A mere list of such bridges would occupy pages. They are generally of shorter span than those mentioned above, rarely exceeding 300 feet, while the number of spans varies from two to seven.

To give here a list of the books which has appeared since Clapeyron's time will also be impossible. We can only indicate two or three which are at the same time valuable and easily accessible :

Bresse: *La mécanique appliquée*, Paris, 1865. (Vol. III contains a tolerably complete mathematical discussion, with tables for facilitating the calculation of moments.)

Winkler: *Theorie der Bruecken*, Vienna, 1872. (Contains the graphical method of Culmann and Mohr, with also analytical investigations.)

Weyrauch: *Theorie der continuirlichen Traeger*, Leipzig, 1873. (A complete analytical discussion of the whole subject. The best work which has yet appeared.)

In our own country have been issued during the past year the works of Greene, Herschel and DuBois, each of which contains valuable contributions to the literature of the subject, and which should be in the library of every student of engineering. On the other hand, works by English authors, which treat of the subject at all, do it in such an imperfect and unsatisfactory manner, that we are forced to consider them as twenty years behind the age.

As the above mentioned works treat only of the theory and calculation of continuous girders, we ought perhaps to say that a book giving an account of the most important continuous and simple, together with suspension and arch bridges, is the admirable descriptive work by Heinzerling, *Die Bruecken in Eisen:* Leipzig, 1870, which is illustrated by over a thousand engravings, and presents a complete history of iron bridge construction.

ERRATA.

Page 15, line 13 ; for l_1 read l.
" 21, " 17 and 18 ; for wl_2 read wl^2.